Skew-Tolerant Circuit Design

Related Titles from Morgan Kaufmann Publishers

Self-Checking and Fault-Tolerant Digital Design
Parag Lala (ISBN 1-12434-370-8)

Readings in Computer Architecture
Mark D. Hill, Norman P. Jouppi, and Gurindar S. Sohi (ISBN 1-55860-539-8)

Logical Effort: Designing Fast CMOS Circuits
Ivan Sutherland, Robert Sproull, and David Harris (ISBN 1-55860-557-6)

The Student's Guide to VHDL
Peter J. Ashenden (ISBN 1-55860-520-7)

The Designer's Guide to VHDL
Peter J. Ashenden (ISBN 1-55860-270-4)

Computer Architecture: A Quantitative Approach, second edition
John L. Hennessy and David A. Patterson (ISBN 1-55860-329-8)

Forthcoming
The Designer's Guide to VHDL, second edition
Peter J. Ashenden (ISBN 1-55860-674-2)

Skew-Tolerant Circuit Design or any of the above titles can be purchased directly from Morgan Kaufmann via the Web at *www.mkp.com* or by phone at 800-745-7323.

Skew-Tolerant Circuit Design

David Harris

Harvey Mudd College

Boz,
Happy hacking!
David Harris

MORGAN KAUFMANN PUBLISHERS

AN IMPRINT OF ACADEMIC PRESS

A Harcourt Science and Technology Company

SAN FRANCISCO   SAN DIEGO   NEW YORK   BOSTON
LONDON   SYDNEY   TOKYO

Senior Editor	Denise E. M. Penrose
Director of Production and Manufacturing	Yonie Overton
Senior Production Editor	Edward Wade
Editorial Assistant	Courtney Garnaas
Cover Design	Ross Carron Design
Text Design	Side By Side Studios, Mark Ong
Composition	Nancy Logan
Cover and Editorial Illustration	Duane Bibby
Technical Illustration	Windfall Software
Copyeditor	Ken DellaPenta
Proofreader	Erin Milnes
Indexer	Steve Rath
Printer	Courier Corporation

Designations used by companies to distinguish their products are often claimed as trademarks or registered trademarks. In all instances where Morgan Kaufmann Publishers is aware of a claim, the product names appear in initial capital or all capital letters. Readers, however, should contact the appropriate companies for more complete information regarding trademarks and registration.

ACADEMIC PRESS
A Harcourt Science and Technology Company
525 B Street, Suite 1900, San Diego, CA 92101-4495, USA
http://www.academicpress.com

Academic Press
24–28 Oval Road, London NW1 7DX, United Kingdom
http://www.hbuk.co.uk/ap/

Morgan Kaufmann Publishers
340 Pine Street, Sixth Floor, San Francisco, CA 94104-3205, USA
http://www.mkp.com

Library of Congress Cataloging-in-Publication Data
Harris, David
    Skew-tolerant circuit design / David Harris.
        p.cm
    Includes bibliographical references and index.
    ISBN 1-55860-636-X
    1. Timing circuits--Design and construction. 2. Integrated circuits--Very large scale integration--Design and construction. 3. Synchronization. I. Title

TK7868.T5 .H37 2001
621.3815--dc21   00-036538

This book has been printed on acid-free paper.

To my parents, Dan and Sally,
who have inspired me to teach

Contents

Preface

As cycle times in high-performance digital systems shrink faster than simple process improvement allows, sequencing overhead consumes an increasing fraction of the clock period. Engineers allocate ever more effort and chip routing resources to the clock network, yet clock skew is becoming a more serious concern. Systems using static logic and edge-triggered flip-flops must budget this clock skew in every clock cycle. Worse yet, aggressive systems attempting to use domino circuits for greater speed budget this clock skew in every half-cycle, or twice in every clock cycle! Moreover, designers have difficulty completely utilizing shrinking clock cycles. The fraction of a gate delay unused at the end of each cycle or half-cycle represents a greater wasted portion of the cycle as the number of gate delays per cycle decreases. All considered, the overhead of traditional domino pipelines can exceed 25% of the cycle time in aggressive systems.

Fortunately, the chip designer can hide much of this overhead through good design techniques. The key to skew-tolerant design is avoiding hard edges in which data must set up before a clock edge but will not continue propagating until after the clock edge. Skew-tolerant static circuits use transparent latches in place of edge-triggered flip-flops to avoid budgeting clock skew and to permit time borrow to balance logic between cycles. Skew-tolerant domino circuits use multiple overlapping clocks to eliminate latches, removing hard edges and hiding the sequencing overhead.

My goal in writing this book is to promote the understanding and appropriate use of skew-tolerant circuit design techniques. Some good designers are comfortable using transparent latches, but many others have exclusively used edge-triggered flip-flops. This book attempts to provide simple and practical explanations of designing with transparent latches to help those who want to switch. Only a smaller cadre of designers are at present skilled with domino circuits. This book seeks to increase the number of domino designers and teach good skew-tolerant domino design practices. The material may be of importance not only to circuit designers, but also to high-speed logic designers who need a good understanding of how their logic will be implemented. It is also important to CAD tool designers. Designers cannot build what their tools cannot analyze so improved CAD tools will be crucial for widespread adoption of skew-tolerant circuit techniques. Of course, the book will also be of use to advanced undergraduate and graduate students interested in chip design.

In contrast to many textbooks and survey articles that merely catalog circuit techniques, this book attempts to evaluate such techniques in the context of present and expected future circuit design environments. Most circuit techniques presented in the literature are impractical or even downright dangerous and are best left to wither in dusty old Ph.D. theses, yet all too often authors do not try to compare the techniques and emphasize the best while warning against the impractical. At the risk of being proven wrong, either now or as technology shifts in the future and changes design trade-offs, this book seeks to judge the value of techniques for industrial designs. Of course, you should not take these judgments on faith, but should continually evaluate the benefits and risks of each circuit for your own applications.

The field of high-speed circuit design changes very rapidly. The historical perspectives section at the end of most chapters describes trends in design, primarily through the 1990s. The perspectives are from my personal experience and literature search. Undoubtedly experienced engineers will have their own points of view; I would welcome hearing about additional information and "stories from the trenches." Send your printable stories to me via ckstories@mkp.com.

Every effort has been made to hunt down and destroy errors, but undoubtedly some still remain. If you discover an error in this book, please contact the publisher by email at ckbugs@mkp.com. The first person to report a technical error will be awarded a $1.00 bounty upon its implementation in future printings of the book. Please check the errata page at *www.mkp.com/clock_skew* to see if a particular bug has already been reported and corrected.

This book is an outgrowth of my Stanford Ph.D. thesis. While laying in a bunk at the Gouter hut on Mont Blanc, I realized that many theses are output-only devices doomed to sit untouched on library shelves. The thesis therefore expanded from a narrow focus on my research about skew-tolerant domino circuits into a broader discussion of skew-tolerant circuit design that integrated best practices of successful designers with my own research. As I have talked and worked with more expert circuit designers, I have found that most of my research is not completely new, but rather has been independently discovered and used by these designers but kept as unpublished trade secrets. However, I hope this book contributes to the field by presenting the ideas of skew-tolerant circuit design in a clear and systematic form.

About the Exercises

In my experience, it is very difficult to learn anything without practice. A number of exercises are available at the end of each chapter for self-study as well as for formal classes. The problems are rated in difficulty on a logarithmic scale, similar to that used by Knuth and Hennessy. A rough guide is listed below. The times assume you have already carefully read and digested the sections of the chapter relevant to the problem. Your mileage may vary.

[10] 1 minute (read and understand)
[20] 15–20 minutes
[30] 2 hours or more (especially if the TV is on)
[50] research problem

Solutions to most even-numbered problems are presented in Appendix B. Please use them wisely; do not turn to the answer until you have made a genuine effort to solve the problems yourself.

Acknowledgments

Many people contributed to the work in this book. My advisor, Mark Horowitz, gave me a framework to think about circuit design and has been a superb critic. He taught me to never accept that a research solution is "good enough" but rather to look for the best possible answer. Ivan Sutherland has been a terrific mentor, giving me the key insight that I needed to break out of writer's block: "A thesis is a document to which three professors will sign their names saying it is. It is well to remember that it is nothing more lest you take too long to finish." Bruce Wooley, on my reading committee, gave the thesis a careful reading and had many useful suggestions. My officemate Ron Ho has been a great help, sounding out technical ideas and saving me many times when I had computer crises. I have enjoyed collaborating with other members of Horowitz's research group and especially thank Jaeha Kim, Dean Liu, Jeff Solomon, Gu Wei, and Evelina Yeung. Zeina Daoud supplied both technical assistance and good conversation.

The book has benefited from the input of several reviewers, including Lynn Conway at the University of Michigan, Tom Fletcher of Intel, Mark Horowitz at Stanford University, Steve Kang at the University of Illinois, Emily J. Shriver of Compaq, Stefano Sidiropolos of RAMBUS, and Gin Yee. The work has been supported through funding from NSF, Stanford, and DARPA. I always learn more when trying to teach, so I would like to thank my students at Stanford, Berkeley, and Harvey Mudd and in industrial courses at Intel, HAL Computer, and Evans & Sutherland.

My first experience with Morgan Kaufmann Publishers [82] was excellent and I have not been disappointed this second time around. Particular recognition goes to Denise Penrose and Edward Wade. I would also like to thank my high school teachers who taught me to write through extensive practice, especially Mrs. Stephens, Mr. Roseth, and Mr. Phillips.

1
Introduction

Long long ago in the Land of Giga Chips the dreaded Dragon of Skew reared his fierce head and snatched away the beautiful Princess Clock. The terrorized villagers summoned Sir Domino and his trusty Squire Static to save the day.

Most digital systems today are constructed using static CMOS logic and edge-triggered flip-flops. Although such techniques have been adequate in the past and will remain adequate in the future for low-performance designs, they will become increasingly inefficient for high-performance components as the number of gates per cycle dwindles and clock skew becomes a greater problem. Designers will therefore need to adopt circuit techniques that can tolerate reasonable amounts of clock skew without an impact on the cycle time. Transparent latches offer a simple solution to the clock skew problem in static CMOS logic. Unfortunately, static CMOS logic is inadequate to meet timing objectives of the highest-performance systems. Therefore, designers turn to domino circuits, which offer greater speed. Unfortunately, traditional domino clocking methodologies [92] lead to circuits that have even greater sensitivity to clock skew and thus can defeat the raw speed advantage of the domino gates. Expert designers of microprocessors and other high-performance systems have long recognized the costs of edge-triggered flip-flops and traditional domino circuits and have used transparent latches and developed a variety of mostly proprietary domino clocking approaches to reduce the overhead. This book formalizes and analyzes skew-tolerant domino circuits, a method of controlling domino gates with multiple overlapping clock phases. Skew-tolerant domino circuits eliminate clock skew from the critical path, hiding the overhead and offering significant performance improvement.

In this chapter, we begin by examining conventional systems built from flip-flops. We see how these systems have overhead that eats into the time available for useful computation. We then examine the trends in throughput and latency for high-performance systems and see that, although the overhead has been modest in the past, flip-flop overhead now consumes a large and increasing portion of the cycle. We turn to transparent latches and show that they can tolerate reasonable amounts of clock skew, reducing the overhead. Next, we examine domino circuits and look at traditional clocking techniques. These techniques have overhead even more severe than that paid by flip-flops. However, by using overlapping clocks and eliminating latches, we find that skew-tolerant domino circuits eliminate all of the overhead. Three case studies illustrate the need for skew-tolerant circuit design.

In Chapter 2, we take a closer look at static CMOS latching techniques, comparing the design and timing of flip-flops, transparent latches, and pulsed latches. We discuss min-delay constraints necessary for correct operation and time borrowing that can help balance logic when used properly. There have been a host of proposed latch designs; we evaluate many of the designs and conclude that the common, simple designs are usually best. For high-performance systems, static CMOS circuits are often too slow, so domino gates are employed. In Chapter 3, we look at skew-tolerant domino design and timing issues. A practical methodology must efficiently combine both static and domino components, so Chapter 4 discusses methodology issues including the static/domino interface, testability, and timing types for high-level verification of proper connectivity. Because we are discussing skew-tolerant circuit design, we are very concerned about the clock waveforms. Chapter 5 explores methods of generating and distributing clocks suitable for skew-tolerant circuits, examines the sources of clock skew in these methods, and describes ways to minimize this skew. Conventional timing analysis tools either cannot handle clock skew or budget it in conservative ways. Chapter 6 describes the problem of timing analysis in skew-tolerant systems and presents simple algorithms for analysis. Finally, Chapter concludes the book by summarizing the main results and greatest future challenges.

1.1 Overhead in Flip-Flop Systems

Most digital systems designed today use positive edge-triggered flip-flops as the basic memory element. A positive edge-triggered flip-flop is often referred to simply as an edge-triggered flip-flop, a D flip-flop, a master-slave flip-flop, or colloquially, just a flop. It has three terminals: input D, clock ϕ, and output Q. When the clock makes a low-to-high transition, the input D is copied to the output Q. The clock-to-Q delay, Δ_{CQ}, is the delay from the rising edge of the clock until the output Q becomes valid. The setup time, Δ_{DC}, is how long the data input D must settle before the clock rises for the correct value to be captured.

Figure 1.1 illustrates part of a static CMOS system using flip-flops. The logic is drawn underneath the clock corresponding to when it operates. Flip-flops straddle the clock edge because input data must set up before the edge and the output becomes valid sometime after the edge. The heavy dashed line at the clock edge represents the cycle boundary. After the flip-flop, data propagates through combinational logic built from

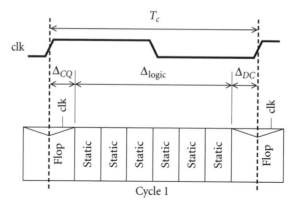

Figure 1.1   Static CMOS system with positive edge-triggered flip-flops

static CMOS gates. Finally, the result is captured by a second flip-flop for use in the next cycle.

How much time is available for useful work in the combinational logic, Δ_{logic}? If the cycle time is T_c, we see that the time available for logic is the cycle time minus the clock-to-Q delay and setup time:

$$\Delta_{\text{logic}} = T_c - \Delta_{CQ} - \Delta_{DC} \tag{1.1}$$

Unfortunately, real systems have imperfect clocks. On account of mismatches in the clock distribution network and other factors that we will examine closely in Chapter 5, the clocks will arrive at different elements at different times. This uncertainty is called clock skew and is represented in Figure 1.2 by a hash of width t_{skew}, indicating a range of possible clock transition times. The bold clock lines indicate the latest possible clocks, which define worst-case timing. Data must set up before the earliest the clock might arrive, yet we cannot guarantee data will be valid until the clock-to-Q delay after the latest clock.

Now we see that the clock skew appears as overhead, reducing the amount of time available for useful work:

$$\Delta_{\text{logic}} = T_c - \Delta_{CQ} - \Delta_{DC} - t_{\text{skew}} \tag{1.2}$$

Flip-flops suffer from yet another form of overhead: imbalanced logic. In an ideal machine, logic would be divided into multiple cycles in such a way that each cycle had exactly the same logic delay. In a real machine, the logic delay is not precisely known at the time cycles are partitioned, so some cycles have more logic and some have less logic. The clock frequency

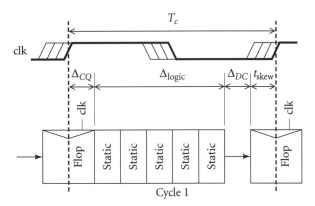

Figure 1.2 Flip-flops including clock skew

must be long enough for the longest cycles to work correctly, meaning excess time in shorter cycles is wasted. The cost of imbalanced logic is difficult to quantify and can be minimized by careful design, but is nevertheless important.

In summary, systems constructed from flip-flops have overhead of the flip-flop delay (Δ_{DC} and Δ_{CQ}), clock skew (t_{skew}), and some amount of imbalanced logic. We will call this total penalty the *sequencing overhead*.[1] In the next section, we will examine trends in system objectives that show sequencing overhead makes up an increasing portion of each cycle.

1.2 Throughput and Latency Trends

Designers judge their circuit performance by two metrics: throughput and latency. *Throughput* is the rate at which data can pass through a circuit; it is related to the clock frequency of the system, so we often discuss cycle time instead. *Latency* is the amount of time for a computation to finish. Simple systems complete the entire computation in one cycle, so latency and throughput are inversely related. Pipelined systems break the computation into multiple cycles called *pipeline stages*. Because each cycle is shorter, new data can be fed to the system more often and the throughput increases. However, because each cycle has some sequencing overhead from flip-flops or other memory elements, the latency of the overall

---

1. We also use the term *clocking overhead*, but such overhead occurs even in asynchronous, unclocked systems, so *sequencing overhead* is a more accurate name.

computation gets longer. For many applications, throughput is the most important metric. However, when one computation is dependent on the result of the previous, the latency of the previous computation may limit throughput because the system must wait until the computation is done. In this section, we will review the relationships among throughput, latency, computation length, cycle time, and overhead. We will then look at the trends in cycle time and find that the impact of overhead is becoming more severe.

When measuring delays, it is often beneficial to use a process-independent unit of delay so that intuition about delay can be carried from one process to another. For example, if I am told that the Hewlett-Packard PA8000 64-bit adder has a delay of 840 ps, I have difficulty guessing how fast an adder of similar architecture would work in my process. However, if I am told that the adder delay is seven times the delay of a fanout-of-4 (FO4) inverter, where an FO4 inverter is an inverter driving four identical copies, I can easily estimate how fast the adder will operate in my process by measuring the delay of an FO4 inverter. Similarly, if I know that microprocessor A runs at 50 MHz in a 1.0-micron process and that microprocessor B runs at 200 MHz in a 0.6-micron process, it is not immediately obvious whether the circuit design of B is more or less aggressive than A. However, if I know that the cycle time of microprocessor A is 40 FO4 inverter delays in its process and that the cycle time of microprocessor B is 25 FO4 inverter delays in its process, I immediately see that B has significantly fewer gate delays per cycle and thus required more careful engineering. The fanout-of-4 inverter is particularly well suited to expressing delays because it is easy to determine, because many designers have a good idea of the FO4 delay of their process, and because the theory of logical effort [82] predicts that cascaded inverters drive a large load fastest when each inverter has a fanout of about 4.

1.2.1   Impact of Overhead on Throughput and Latency

Suppose a computation involves a total combinational logic delay X. If the computation is pipelined into N stages, each stage has a logic delay $\Delta_{\text{logic}} = X/N$. As we have seen in the previous section, the cycle time is the sum of the logic delay and the sequencing overhead:

$$T_c = \frac{X}{N} + \text{overhead} \qquad (1.3)$$

The latency of the computation is the sum of the logic delay and the total overhead of all N stages:

$$\text{latency} = NT_c = X + N \bullet \text{overhead} \tag{1.4}$$

Equations 1.3 and 1.4 show how the impact of a fixed overhead increases as a computation is pipelined into more stages of shorter length. The overhead becomes a greater portion of the cycle time T_c, so less of the cycle is used for useful computation. Moreover, the latency of the computation actually increases with the number of pipe stages N because of the overhead. Because latency matters for some computations, system performance can actually decrease as the number of pipe stages increases.

EXAMPLE 1.1   Consider a system built from static CMOS logic and flip-flops. Let the setup (Δ_{DC}) and clock-to-Q (Δ_{CQ}) delays of the flip-flop be 1.5 FO4 inverter delays. Suppose the clock skew (t_{skew}) is 2 FO4 inverter delays. What percentage of the cycle is wasted by sequencing overhead if the cycle time T_c is 40 FO4 delays? 24 FO4 delays? 16 TFO4 delays? 12 FO4 delays?

SOLUTION   The sequencing overhead is $1.5 + 1.5 + 2 = 5$ FO4 delays. The percentage of the cycle consumed by overhead is shown in Table 1.1. This example shows that although the sequencing overhead was small as a percentage of cycle time when cycles were long, it becomes very severe as cycle times shrink.   ∎

Table 1.1   Sequencing overhead

Cycle time	Percentage overhead
40	13
24	21
16	31
12	42

The exponential increase in microprocessor performance, doubling about every 18 months [36], has been caused by two factors: better microarchitectures that increase the average number of instructions executed per cycle (IPC), and shorter cycle times. The cycle time improvement is a combination of steadily improving transistor performance and better circuit design using fewer gate delays per cycle. To evaluate the importance of sequencing overhead, we must tease apart these elements of performance increase to identify the trends in gates per cycle. Let us look both at the historical trends of Intel microprocessors and at industry predictions for the future.

1.2.2   Historical Trends

Figure 1.3 shows a plot of Intel microprocessor performance from the 16 MHz 80386 introduced in 1985 to the 800 MHz Pentium II processors selling in 1999 [42, 57]. The performance has increased at an incredible 53% per year, thus doubling every 19.5 months. This exponential increase in processor performance is often called "Moore's law" [59], although technically Gordon Moore's original predictions only referred to the exponential growth of transistors per integrated circuit, not the performance growth.

Figure 1.4 shows a plot of the processor clock frequencies, increasing at a rate of 30% per year. Some of this increase comes from faster transistors, and some comes from using fewer gates per cycle.

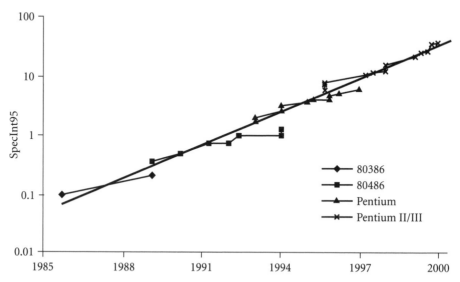

Figure 1.3   Intel microprocessor performance
Note: Performance is measured in SpecInt95. For processors before the
90 MHz Pentium, SpecInt95 is estimated from published MIPS data
with the conversion 1 MIPS = 0.0185 SpecInt95.

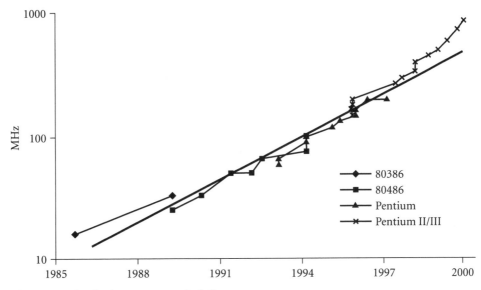

Figure 1.4   Intel microprocessor clock frequency

Because we are particularly interested in the number of FO4 inverters per cycle, we need to estimate how FO4 delay improves as transistors shrink. Figure 1.5 shows the FO4 inverter delay of various MOSIS processes over many years. The delays are linearly scaling with the feature size and, averaging across voltages commonly used at each feature size, are well fit by the equation

$$\text{FO4 delay} = 475f \tag{1.5}$$

where f is the minimum drawn channel length measured in microns and delay is measured in picoseconds.

Using this delay model and data about the process used in each part, we can determine the number of FO4 delays in each cycle, shown in Figure 1.6. Notice that for a particular product, the number of gate delays in a cycle is initially high and gradually decreases as engineers tune critical paths in subsequent revisions on the same process and jumps as the chip is compacted to a new process that requires retuning. Overall, the number of FO4 delays per cycle has decreased at 12% per year.

Putting everything together, we find that the 1.53 times annual historical performance increase can be attributed to 1.17 times from microarchitectural improvements, 1.17 times from process improvements, and 1.12 times from fewer gate delays per cycle.

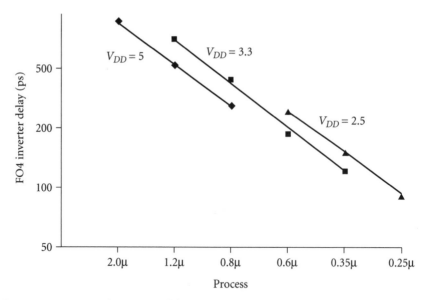

Figure 1.5   Fanout-of-4 inverter delay trends

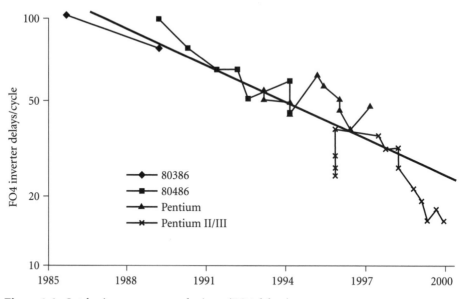

Figure 1.6   Intel microprocessor cycle times (FO4 delays)

1.2.3   Future Predictions

The Semiconductor Industry Association (SIA) issued a roadmap in 1997
[71] predicting the evolution of semiconductor technology over the next

Table 1.2   SIA roadmap of clock frequencies

Process (μm)	Year	Frequency (MHz)	Cycle time (FO4)
0.25	1997	750 (600)	13.3 (16.7)
0.18	1999	1250 (733)	11.1 (18.9)
0.15	2001	1500	11.1
0.13	2003	2100	9.2
0.10	2006	3500	7.1
0.07	2009	6000	6.0
0.05	2012	10,000	5.0

15 years. Although such predictions are always fraught with peril, let us look at what the predictions imply for cycle times in the future.

Table 1.2 lists the year of introduction and estimated local clock frequencies predicted by the SIA for high-end microprocessors in various processes. The SIA assumes that future chips may have very fast local clocks serving small regions, but will use slower clocks when communicating across the die. The table also contains the predicted FO4 delays per cycle using a formula that

$$FO4 \text{ delay} = 400f \qquad (1.6)$$

which better matches the delay of 0.25- and 0.18-micron processes than does Equation 1.5; these current processes are a more appropriate operating point to linearize around when predicting the future. The predictions proved somewhat optimistic in 1997 and 1999.

This roadmap shows a 7% annual reduction in cycle time. The predicted cycle time of only 5 FO4 inverter delays in a 0.05-micron process seems at the very shortest end of feasibility because it is nearly impossible for a loaded clock to swing rail to rail in such a short time. Nevertheless, it is clear that sequencing overhead of flip-flops will become an unacceptable portion of the cycle time.

1.2.4   Conclusions

In summary, we have seen that sequencing overhead was negligible in the 1980s when cycle times were nearly 100 FO4 delays. As cycle times measured in gate delays continue to shrink, the overhead becomes more important and is now a major and growing obstacle for the design of high-performance systems. We have not discussed clock skew in this section, but we will see in Chapter 5 that clock skew, as measured in gate delays, is likely to grow in future processes, making that component of overhead even worse. Clearly, flip-flops are becoming unacceptable, and we need to use better design methods that tolerate clock skew without introducing overhead. In the next section, we will see how transparent latches accomplish exactly that.

1.3   Skew-Tolerant Static Circuits

We can avoid the clock skew penalties of flip-flops by instead building systems from two-phase transparent latches, as has been done since the early days of CMOS [56]. Transparent latches have the same terminals as flip-flops: data input D, clock ϕ, and data output Q. When the clock is high, the latch is transparent and the data at the input D propagates through to the output Q. When the clock is low, the latch is opaque and the output retains the value it last had when transparent. Transparent latches have three important delays. The clock-to-Q delay, Δ_{CQ}, is the time from when the clock rises until data reaches the output. The D-to-Q delay, Δ_{DQ}, is the time from when new data arrives at the input while the latch is transparent until the data reaches the output. Δ_{CQ} is typically somewhat longer than Δ_{DQ}. The setup time, Δ_{DC}, is how long the data input D must settle before the clock falls for the correct value to be captured.

Figure 1.7 illustrates part of a static CMOS system using a pair of transparent latches in each cycle. One latch is controlled by clk, while the other is controlled by its complement clk_b. In this example, we show the data arriving at each latch midway through its half-cycle. Therefore, each latch is transparent when its input arrives and incurs only a D-to-Q delay rather than a clock-to-Q delay. Because data arrives well before the falling edge of the clock, setup times are trivially satisfied.

How much time is available for useful work in the combinational logic, Δ_{logic}? If the cycle time is T_c, we see that

$$\Delta_{\text{logic}} = T_c - 2\Delta_{DQ} \tag{1.7}$$

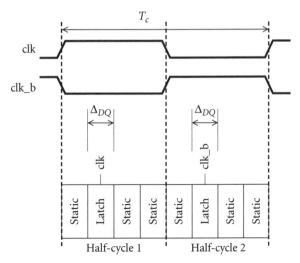

Figure 1.7   Static CMOS system with transparent latches

We will see in Chapter 2 that flip-flops are built from pairs of back-to-back latches so that the time available for logic in systems with no skew is about the same for flip-flops and transparent latches. However, transparent-latch systems can tolerate clock skew without cycle time penalty, as seen in Figure 1.8. Although the clock waveforms have some uncertainty from skew, the clock is certain to be high when data arrives at the latch so the data can propagate through the latch with no extra overhead. Data still arrives well before the earliest possible skewed clock edge, so setup times are still satisfied.

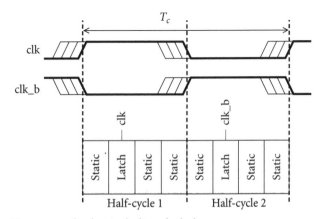

Figure 1.8   Transparent latches including clock skew

Finally, static latches avoid the problem of imbalanced logic through a phenomenon called *time borrowing*, also known as *cycle stealing* by engineers at Big Blue. We see from Figure 1.8 that each latch can be placed in any of a wide range of locations in its half-cycle and still be transparent when the data arrives. This means that not all half-cycles need to have the same amount of static logic. Some can have more and some can have less, meaning that data arrives at the latch later or earlier without wasting time as long as the latch is transparent at the arrival time. Hence, if the pipeline is not perfectly balanced, a longer cycle may borrow time from a shorter cycle so that the required clock period is the average of the two rather than the longer value. In Chapter 2 we will quantify how much time borrowing is possible.

In summary, systems constructed from transparent latches still have overhead from the latch propagation delay (Δ_{DQ}) but eliminate the overhead from reasonable amounts of clock skew and imbalanced logic. This improvement is especially important as cycle times decrease, justifying a switch to transparent latches for high-performance systems.

1.4   Domino Circuits

To construct systems with fewer gate delays per cycle, designers may invent more efficient ways to implement particular functions or may use faster gates. The increasing transistor budgets allow parallel structures that are faster; for example, adders progressed from compact but slow ripple carry architectures to larger carry look-ahead designs to very fast but complex tree structures. However, there is a limit to the benefits from using more transistors, so designers are increasingly interested in faster circuit families, in particular domino circuits. Domino circuits are constructed from alternating dynamic and static gates. In this section, we will examine how domino gates work and see why they are faster than static gates. Gates do not exist in a vacuum; they must be organized into pipeline stages. When domino circuits are pipelined in the same way that two-phase static circuits have traditionally been pipelined, they incur a great deal of sequencing overhead from latch delay, clock skew, and imbalanced logic. By using overlapping clocks and eliminating the latches, we will see that skew-tolerant domino circuits can hide this overhead to achieve dramatic speedups.

1.4.1 Domino Gate Operation

To understand the benefits of domino gates, we will begin by analyzing the delay of a gate. Remember that the time to charge a capacitor is

$$\Delta t = \frac{C}{I} \Delta V \tag{1.8}$$

For now we will just consider gates that swing rail to rail, so ΔV is V_{DD}. If a gate drives an identical gate, the load capacitance and input capacitance are equal (neglecting parasitics), so it is reasonable to consider the C/I ratio of the gate's input capacitance to the current delivered by the gate as a metric of the gate's speed. This ratio is called the *logical effort* [82] of the gate and is normalized to one for a static CMOS inverter. It is higher for more complex static CMOS gates because series transistors in complex gates must be larger and thus have more input capacitance to deliver the same output current as an inverter.

Static circuits are slow because inputs must drive both NMOS and PMOS transistors. Only one of the two transistors is on, meaning that the capacitance of the other transistor loads the input without increasing the current drive of the gate. Moreover, the PMOS transistor must be particularly large because of its poor carrier mobility and thus adds much capacitance.

A dynamic gate replaces the large, slow PMOS transistors of a static CMOS gate with a single clocked PMOS transistor that does not load the input. Figure 1.9 compares static and dynamic NOR gates. The dynamic gates operate in two phases: precharge and evaluation. During the precharge phase, the clock is low, turning on the PMOS device and pulling the

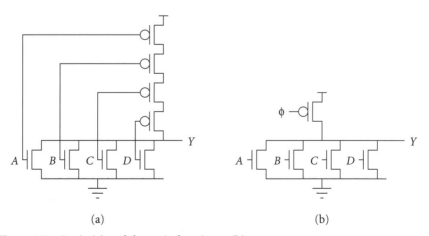

(a)   (b)

Figure 1.9 Static (a) and dynamic four-input (b) NOR gates

output high. During the evaluation phase, the clock is high, turning off the PMOS device. The output of the gate may *evaluate* low through the NMOS transistor stack.

The dynamic gate is faster than the static gate for two reasons. One is the greatly reduced input capacitance. Another is the fact that the dynamic gate output begins switching when the input reaches the transistor threshold voltage, V_t. This is sooner than the static gate output, which begins switching when the input passes roughly $V_{DD}/2$. This improved speed comes at a cost, however: dynamic gates must obey precharge and monotonicity rules, are more sensitive to noise, and dissipate more power because of their higher activity factors.

The *precharge* rule says that there must be no active path from the output to ground of a dynamic gate during precharge. If this rule is violated, there will be contention between the PMOS precharge transistor and the NMOS transistors pulling to ground, consuming excess power and leaving the output at an indeterminate value. Sometimes the precharge rule can be satisfied by guaranteeing that some inputs are low. For example, in the four-input NOR gate, all four inputs must be low during precharge. In a four-input NAND gate, if any input is low during precharge, there will be no contention. It is commonly not possible to guarantee inputs are low, so often an extra *clocked evaluation transistor* is placed at the bottom of the dynamic pulldown stack, as shown in Figure 1.10. Gates with and without clocked evaluation transistors are sometimes called *footed* and *unfooted* [62]. Unfooted gates are faster but require more complex clocking to prevent both PMOS and NMOS paths from being simultaneously active.

The *monotonicity* rule states that all inputs to dynamic gates must make only low-to-high transitions during evaluation. Figure 1.11 shows a circuit that violates the monotonicity rule and obtains incorrect results. The circuit consists of two cascaded dynamic NOR gates. The first computes

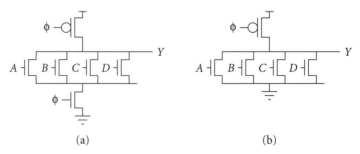

(a)   (b)

Figure 1.10   Footed (a) and unfooted four-input dynamic (b) NOR gates

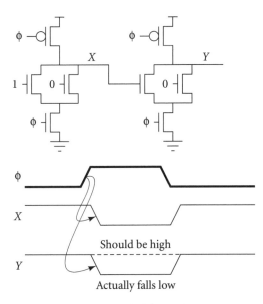

Figure 1.11 Incorrect operation of cascaded dynamic gates

$X = \text{NOR}(1, 0) = 0$. The second computes $Y = \text{NOR}(X, 0)$, which should be 1. Node X is initially high and falls as the first NOR gate evaluates. Unfortunately, the second NOR gate sees that input X is high when ϕ rises and thus pulls down output Y incorrectly. Because the dynamic NOR gate has no PMOS transistors connected to the input, it cannot pull Y back high when the correct value of X arrives, so the circuit produces an incorrect result. The problem occurred because X violated the monotonicity rule by making a high-to-low transition while the second gate is in evaluation.

It is not possible to cascade dynamic gates directly with the same clock without violating the monotonicity rule because each dynamic output makes a high-to-low transition during evaluation while dynamic inputs require low-to-high transitions during evaluation. An easy way to solve the problem is to insert an inverting static gate between dynamic gates, as shown in Figure 1.12. The dynamic/static gate pair is called a *domino gate,* which is slightly misleading because it is actually two gates. A cascade of domino gates precharge simultaneously like dominos being set up. During evaluation, the first dynamic gate falls, causing the static gate to rise, the next dynamic gate to fall, and so on like a chain of dominos toppling.

Unfortunately, to satisfy monotonicity we have constructed a pair of OR gates rather than a pair of NOR gates. In Chapter 3 we will return to the monotonicity issue and see how to implement arbitrary functions with domino gates.

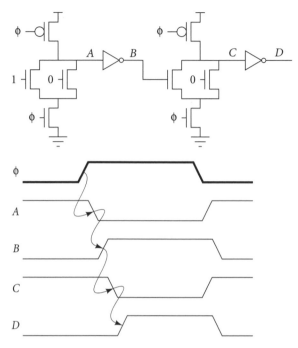

Figure 1.12 Correct operation with domino gates

Mixing static gates with dynamic gates sacrifices some of the raw speed offered by the dynamic gate. We can regain some of this performance by using *HI-skew*[2] static gates with wider-than-usual PMOS transistors [82] to speed the critical rising output during evaluation. Moreover, the static gates may perform arbitrary functions rather than being just inverters [87]. All considered, domino logic runs 1.5 to 2 times faster than static CMOS logic [49] and is therefore attractive enough for high-speed designs to justify its extra complexity.

1.4.2 Traditional Domino Clocking

After domino gates evaluate, they must be precharged before they can be used in the next cycle. If all domino gates were to precharge simultaneously, the circuit would waste time while only precharging, not useful computation, takes place. Therefore, domino logic is conventionally divided into two phases or half-cycles, ping-ponged such that the first

---

2. Don't confuse the word *skew* in "HI-skew" gates with "clock skew."

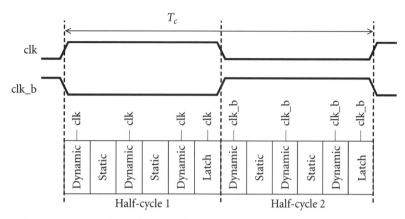

Figure 1.13 Traditional domino clocking scheme

phase evaluates while the second precharges, and then the first phase precharges while the second evaluates. In a traditional domino clocking scheme [92], latches are used between phases to sample and hold the result before it is lost to precharge, as illustrated in Figure 1.13. The scheme appears very similar to the use of static logic and transparent latches discussed in the previous section. Unfortunately, we will see that such a scheme has enormous sequencing overhead.

With ideal clocks, the first dynamic gate begins evaluating as soon as the clock rises. Its result ripples through subsequent gates and must arrive at the latch a setup time before the clock falls. The result propagates through the latch, so the overhead of each latch is the maximum of its setup time and D-to-Q propagation delay. The latter time is generally larger, so the total time available for computation in the cycle is

$$\Delta_{\text{logic}} = T_c - 2\Delta_{DQ} \qquad (1.9)$$

Unfortunately, a real pipeline like that shown in Figure 1.14 experiences clock skew. In the worst case, the dynamic gate and latch may have greatly skewed clocks. Therefore, the dynamic gate may not begin evaluation until the latest skewed clock, while the latch must set up before the earliest skewed clock. Hence, clock skew must be subtracted not just from each cycle, as it was in the case of a flip-flop, but from each half-cycle! Assuming the sum of clock skew and setup time are greater than the latch D-to-Q delay, the time available for useful computation becomes

$$\Delta_{\text{logic}} = T_c - 2\Delta_{DC} - 2t_{\text{skew}} \qquad (1.10)$$

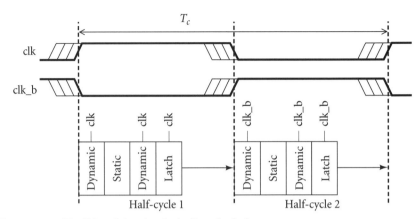

Figure 1.14   Traditional domino including clock skew

As with flip-flops, traditional domino pipelines also suffer from imbalanced logic. In summary, traditional domino circuits are slow because they pay overhead for latch delay, clock skew, and imbalanced logic. In the case study of the Alpha 21164 microprocessor in Section 1.5.2, we will see that this overhead can easily reach a quarter of the cycle time.

1.4.3   Skew-Tolerant Domino

Both flip-flops and traditional domino circuits launch data on one edge and sample it on another. These edges are called *hard edges* or *synchronization points* because the arrival of the clock determines the exact timing of data. Even if data is available early, the hard edges prevent subsequent stages from beginning early. Static CMOS pipelines with transparent latches avoided the hard edges and therefore could tolerate some clock skew and use time borrowing to compensate for imbalanced logic. Some domino designers have recognized that this fundamental idea of softening the hard clock edges can be applied to domino circuits as well. Although a variety of schemes were invented at most microprocessor companies in the mid-1990s (e.g., [41]), the schemes have generally been held as trade secrets. This section explains how such *skew-tolerant domino* circuits operate. In Chapters 3 and 5 we will return to more subtle choices in the design and clocking of such circuits.

The basic problem with traditional domino circuits is that data must arrive by the end of one half-cycle but will not depart until the beginning of the next half-cycle. Therefore, the circuits must budget skew between the clocks and cannot borrow time. We can overcome this problem by using overlapping clocks, as shown in Figure 1.15. This figure shows a skew-tolerant domino clocking scheme with two overlapping clock phases. Instead of using clk and its complement, we now use overlapping clocks ϕ_1 and ϕ_2. We partition the logic into *phases* instead of half-cycles because in general we will allow more than two overlapping phases. The clocks overlap enough that even under worst-case clock skews providing minimum overlap, the first gate B in the second phase has time to evaluate before the last gate A in the first phase begins precharge. As with static latches, the gates are guaranteed to be ready to operate when the data arrives even if skews cause modest variation in the arrival time of the clock. Therefore we do not need to budget clock skew in the cycle time.

Another advantage of skew-tolerant domino circuits is that latches are not necessary within the domino pipeline. We ordinarily need latches to hold the result of the first phase's computation for use by the second phase when the first phase precharges. In skew-tolerant domino, the overlapping clocks insure that the first gate in the second phase has enough time to evaluate before ϕ_1 falls and the first phase begins precharge. When the first phase precharges, the dynamic gates will pull high and therefore the static gates will fall low. This means that the input to the second phase falls low, seemingly violating the monotonicity rule that inputs to dynamic gates must make only low-to-high transitions while the gates are in evaluation. What are the consequences of violating monotonicity? Gate B will remain at whatever value it evaluated to based on the results of the

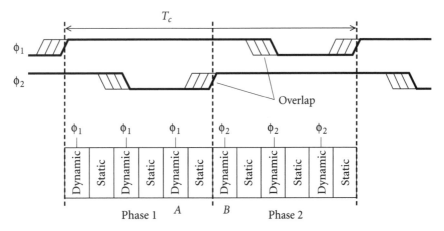

Figure 1.15 Two-phase skew-tolerant domino circuits

first half-cycle when its inputs fall low because both the pulldown transistors and the precharge transistor will be off. This is exactly what we want:
gate B will keep its value even when Phase 1 precharges. Hence, there is no
need for a latch at the end of Phase 1 to remember the result during precharge. The entire cycle is available for useful computation; there is no
sequencing overhead from latch delay or clock skew:

$$\Delta_{\text{logic}} = T_c \qquad\qquad (1.11)$$

Finally, skew-tolerant domino circuits can allow time borrowing if the
overlap between clock phases is larger than the clock skew. The guaranteed overlap is the nominal overlap minus uncertainty due to the clock
skew. Gates in either Phase 1 or Phase 2 may evaluate during the overlap
period, allowing time borrowing by letting gates that nominally evaluate
during Phase 1 to run late into the second phase. As usual, it is hard to
quantify the benefits of time borrowing, but it is clear that the designer
has greater flexibility.

In summary, skew-tolerant domino circuits use overlapping clocks to
eliminate latches and remove all three sources of overhead that plague
traditional domino circuits: clock skew, latch delay, and imbalanced logic.
We will see in Section 1.5.2 that this overhead can be about 25% of the
cycle time of a well-designed system today and will increase as cycles get
shorter. Therefore, skew-tolerant domino is significantly faster than traditional domino in aggressive systems.

Two-phase skew-tolerant domino circuits conveniently illustrate the
benefits of skew-tolerant domino design, but prove to suffer from hold
time problems and offer limited amounts of time borrowing and skew
tolerance. Chapter 3 generalizes the idea to multiple overlapping clock
phases and derives formulae describing the available time borrowing and
skew tolerance as a function of the number of phases and the duty cycle.
Chapter 5 addresses the challenges of producing the multiple overlapping
phases.

1.5   Case Studies

To illustrate the benefits of skew-tolerant circuit design, let us consider
three brief case studies. The first is an application-specific integrated circuit (ASIC) using static logic and edge-triggered flip-flops in which
sequencing overhead got out of hand. The second is the Alpha 21164, a

much more aggressive and heavily optimized design in which traditional domino circuits still impose an overhead of about 25% of the cycle time. The third is a study of timing analysis showing how better analysis leads to less pessimistic results. The first two studies come from real industrial projects, while the third is contrived to illustrate the idea in a simple way.

1.5.1   Sequencing Overhead in a Static ASIC

A colleague spent the summer of 1998 designing an application-specific integrated circuit targeting 200 MHz operation in the TSMC 0.25-micron process. Names are withheld to protect the innocent. The process provided an estimated 120 ps fanout-of-4 inverter delay, so the 5 ns cycle time was about 40 FO4 inverter delays. This is reasonably aggressive for a static CMOS ASIC design, but is not unreasonable with careful attention to pipeline partitioning and circuit implementation.

Unfortunately, as is often the case at the forefront of technology, the process was new and untested. As a result, the clock skew budget was set to be ±500 ps. In the notation of this book, $t_{skew} = 1000$ ps because the launching clock might be skewed late while the receiving clock is skewed early. The cell library included an edge-triggered flip-flop with a total delay of $\Delta_{DC} + \Delta_{CQ} = 400$ ps. With these figures, the sequencing overhead consumed 1.4 ns, or 28% of the cycle time, neglecting the further overhead of imbalanced logic. Only 3.6 ns, or 30 FO4 inverter delays, were left for useful logic. Thus the design became much more challenging than it should have been. The conservative estimate of clock skew led to higher engineering costs as the engineer was forced to overdesign the critical paths.

Several approaches could have eased the design. The most effective would have been to use latches rather than flip-flops on the critical paths. This would have removed clock skew from the cycle time budget, relinquishing about 8 FO4 inverter delays for useful computation. Unfortunately, the ASIC house did not have experience with, or tools for, latch-based design, so this was not an option at the time. However, latch-based design has been practiced in the industry for decades and is supported by many CAD packages, so an investment in such capability would provide payoffs on many future high-speed designs.

Another approach would have been to understand the sources and magnitude of the clock skew better. Indeed, late in the design, the skew budget was reduced to 720 ps through better analysis. This offered little solace to the designer who had spent enormous amounts of effort optimizing paths to operate under the original skew budget. Most paths communicate between nearby flip-flops that see much less skew than flip-flops on opposite corners of the die. If such information were made explicit to the analysis tools, the overhead could also be reduced on those paths. We'll look at this idea more closely in Section 1.5.3 and Chapter 6.

1.5.2   Sequencing Overhead in the Alpha 21164

The Alpha 21164 was the fastest commercial microprocessor of its day, achieving 500 MHz in 1996 using a 0.35-micron process years before most of its competition caught up using 0.18-micron processes. It relied on extensive use of domino circuits to achieve its speed. Traditional two-phase domino was most widely used [26].

We can estimate the sequencing overhead of the domino given a reported clock skew of 200 ps. If we assume the latch setup time is only 50 ps, we find a total of 250 ps of overhead must be budgeted in each half-cycle and 500 ps must be budgeted per cycle. This accounts for 25% of the 2 ns cycle time, neglecting any overhead from imbalanced logic.

The Alpha designers were unwilling to sacrifice this much time to overhead. They used two approaches to reduce the overhead. One was to use overlapping clocks in the ALU self-bypass path to eliminate one of the latches, effectively acting as two-phase skew-tolerant domino. A second was to compute local skews between nearby clocked elements and use these smaller skews wherever possible rather than budgeting global skew everywhere. Data circulating in the ALU self-bypass loop sees only this local skew. At the time of design, standard timing analyzers could not handle different amounts of skew between different latches, so the design team had to use an internal timing analyzer. Now this capability is becoming available in major commercial tools such as Pathmill with the Clock Skew Option.

1.5.3   Timing Analysis with Clock Skew

When clocks experience different amounts of clock skew relative to one another, a good timing analyzer will determine the clocks that launched and terminated a path and allocate only the necessary skew rather than

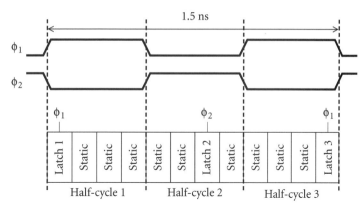

Figure 1.16   Path with time borrowing for timing analysis

pessimistically budgeting the worst-case skew. To see the difference, consider the path in Figure 1.16 targeting 1 GHz operation. In this path, let us assume that ϕ_1 has only 10 ps of skew relative to other copies of itself, but that ϕ_2 may experience 200 ps of skew relative to ϕ_1. Assume a setup time of 60 ps and a D-to-Q delay of 50 ps for each latch.

Suppose the path from latch 1 to latch 2 is long and borrows 210 ps into half-cycle 2, as shown in the figure. The path from latch 2 to latch 3 is also long. We would like our timing analyzer to tell us if the path will violate the setup time at latch 3. How much clock skew must be budgeted in this path?

At first, we might assume that because there is 200 ps of skew between ϕ_2 and ϕ_1, the data must arrive at latch 3 at a setup time and clock skew before the falling edge of ϕ_2, that is, no later than 1.5 ns – 0.2 ns – 0.06 ns = 1.24 ns. Indeed, at least one commercial timing analyzer reports exactly this.

Looking more closely, we realize that the data arrives at latch 2 while latch 2 is transparent. Therefore, latch 2 does not retime the data departure. The path is launched from latch 1 and must budget only 10 ps of skew at latch 3 because the launching and receiving clocks are both ϕ_1. The true required time at latch 3 is 1.5 ns – 0.01 ns – 0.06 ns = 1.43 ns. By considering the launching and receiving clocks, we determine that we have 190 ps more time available than we might have pessimistically assumed.

Figure 1.17 shows a sample of a report we would like to receive from timing analysis. In this example, we assume the logic delay between latch

```
Startpoint: latch2/reg (positive level-sensitive latch clocked by phi2)
Endpoint: latch3/reg (positive level-sensitive latch clocked by phi1)

Point                                              Incr      Path
--------------------------------------------------------------------
clock phi2 (rise edge)                             0.50      0.50
clock network delay (ideal)                        0.00      0.50
time given to startpoint (relative to phi1)        0.21      0.71
latch2/reg/D (LD1QCFP)                             0.00      0.71 r
latch2/reg/Q (LD1QCFP)                             0.05      0.76
...
latch3/reg/D (LD1QAFP)                             0.60      1.36 r
data arrival time                                            1.36
clock phi1 (rise edge)                             1.00      1.00
clock network delay (ideal)                        0.00      1.00
latch1/reg/G (LD1QAFP)                                       1.00 r
time borrowed from endpoint                        0.36      1.36
data required time                                           1.36
--------------------------------------------------------------------
data required time                                           1.36
data arrival time                                            1.36
--------------------------------------------------------------------
slack (MET)                                                  0.00
Time Borrowing Information
--------------------------------------------------------------------
phi1 pulse width                                             0.50
library setup time                                          -0.06
inter-clock uncertainty (phi1, phi1)                        -0.01
--------------------------------------------------------------------
max time borrow                                              0.43
actual time borrow                                           0.36
--------------------------------------------------------------------
```

Figure 1.17 Timing analysis report

2 and latch 3 is 0.6 ns. The report indicates that data departs latch 2 at 0.71 ns relative to ϕ_1. It has a 0.05 ns propagation delay through the latch and 0.60 ns delay through the logic, arriving at latch 3 at 1.36 ns. This requires borrowing 0.36 ns into half-cycle 3. Because the path was launched by ϕ_1 and received by ϕ_1, only 0.01 ns of skew must be budgeted. The maximum time available for borrowing is 0.43 ns, determined by the half-cycle time less the setup and appropriate clock skew. Therefore the path meets the setup time with an acceptable amount of time borrowing. If 0.2 ns of skew were budgeted, the path would miss timing by 120 ps.

Timing analysis with different amounts of clock skew between different clocks in systems supporting time borrowing is somewhat subtle, as this example has illustrated. The concept of departure times relative to various clocks is developed in Chapter 6, along with algorithms to perform the timing analysis. Be aware that your timing analyzer may not yet support such analysis.

1.6   A Look Ahead

In this chapter we have examined the sources of sequencing overhead and seen that it is a growing problem in high-speed digital circuits, as summarized in Table 1.3. While flip-flops and traditional domino circuits have severe overhead, transparent latches and skew-tolerant domino circuits hide clock skew and allow time borrowing to balance logic between pipeline stages. In the subsequent chapters we will flush out these ideas to present a complete methodology for skew-tolerant circuit design of static and dynamic circuits.

Chapter 2 focuses on static circuit design. It examines the three commonly used memory elements: flip-flops, transparent latches, and pulsed latches. While flip-flops clearly have the worst overhead, transparent latches and pulsed latches each have pros and cons. Pulsed latches are faster in an ideal environment, but transparent latches have less critical race conditions and can tolerate more clock skew and time borrowing. We take a closer look at time borrowing and latch placement, then consider hold time violations, which we have ignored in this introduction. Finally, we survey a variety of memory element implementations.

Chapter 3 moves on to domino circuit design. It addresses the question of how to best clock skew-tolerant domino circuits and derives how much skew and time borrowing can be handled as a function of the

Table 1.3   Sequencing overhead

Sequencing method	Sequencing overhead
Flip-flops	$\Delta_{CQ} + \Delta_{DC} + t_{\text{skew}} +$ imbalanced logic
Transparent latches	$2\Delta_{DC}$
Traditional domino	$2\Delta_{DC} + 2t_{\text{skew}} +$ imbalanced logic
Skew-tolerant domino	0

number of clock phases and their duty cycles. Given these formulae, hold times, and practical clock generation issues, we conclude that four-phase skew-tolerant domino circuits are a good way to build systems. We then return to general domino design issues, including monotonicity, footed and unfooted dynamic gates, and noise.

Chapter 4 puts together static and domino circuits into a coherent skew-tolerant circuit design methodology. It looks at the interface between the two circuit families and shows that the static-to-domino interface must budget clock skew, motivating the designer to build critical rings entirely in domino for maximum performance. It describes the use of timing types to verify proper connectivity in static circuits, then extends timing types to handle skew-tolerant domino. Finally, it addresses issues of testability and shows that scan techniques can serve both latches and skew-tolerant domino in a simple and elegant way.

None of these skew-tolerant circuit techniques would be practical if providing the necessary clocks introduced more skew than the techniques could handle. Chapter 5 addresses clock generation and distribution. Many experienced designers reflexively cringe when they hear schemes involving multiple clocks because it is virtually impossible to route more than one high-speed clock around a chip with acceptable skew. Instead, we distribute a single clock across the chip and locally produce the necessary phases with the final-stage clock drivers. We analyze the skews from these final drivers and conclude that although the delay variation is non-negligible, skew-tolerant circuits are on the whole a benefit. In addition to tolerating clock skew, good systems minimize the skew that impacts each path. By considering the components of clock skew and dividing a large die into multiple clock domains, we can budget smaller amounts of skew in most paths than we must budget across the entire die.

By this point, we have developed all the ideas necessary to build fast skew-tolerant circuits. With a little practice, skew-tolerant circuit design is no harder than conventional techniques. However, it is impossible to build multimillion transistor ICs unless we have tools that can analyze and verify our circuits. In particular, we need to be able to check if our circuits can meet timing objectives given the actual skews between clocks in various domains. Chapter 6 addresses this problem of timing analysis, extending previous formulations to handle multiple domains of clock skew.

By the end of this book, you should have a thorough understanding of how to design skew-tolerant static and domino circuits. Although such design is not difficult, it has been ignored by the bulk of engineers and computer-aided design systems because flip-flops were adequate when cycle times were long and thus sequencing overhead was small. High-end microprocessors push circuit performance to the limit and will benefit from skew-tolerant domino circuits to reduce overhead. Application-specific integrated circuits will generally have less aggressive frequency targets and are unlikely to employ domino circuits until signal integrity tools improve. Nevertheless, many ASICs will be fast enough that flip-flop overhead becomes significant, and a switch to skew-tolerant latches may make design easier. Given these trends, skew-tolerant circuit design should be an exciting area in the coming years.

1.7   Exercises

[15]   **1.1**   You are building the AlphaNot microprocessor and plan to use flip-flops as your state elements. Suppose the setup and clock-to-Q delays of the flip-flops are 1.5 FO4 delays each. Assume there is no clock skew. You are targeting a 0.18-micron process with a 60 ps FO4 inverter delay. If the operating frequency is to be 600 MHz, what fraction of the cycle time is wasted for sequencing overhead? Repeat if the operating frequency is 1 GHz.

[15]   **1.2**   Repeat Exercise 1.1 if the clock skew is 150 ps.

[15]   **1.3**   Repeat Exercise 1.1 using transparent latches instead of flip-flops. Your latches have setup and clock-to-Q delays of 1.5 FO4 delays each. They have a D-to-Q delay of 1.3 FO4 delays.

[15]   **1.4**   Repeat Exercise 1.3 if the clock skew is 150 ps.

[15]   **1.5**   You are designing an IEEE single-precision floating-point multiplier targeting 200 MHz operation in a 0.18-micron process using synthesized static CMOS logic and conventional flip-flops. The flip-flops in your cell

library have a setup time of 200 ps and a clock-to-Q delay of 300 ps for the expected loading. You are budgeting clock skew of 400 ps. How much time (in nanoseconds) is available for useful multiplier logic?

[15]   **1.6**   Repeat Exercise 1.5 if you are using IBM's 0.16-micron SA-27 process using synthesized static CMOS logic and conventional flip-flops. An excerpt from IBM's cell library data book is shown in Figure 1.18. Assume you will use the LDE0001 flip-flop (ignore the A, B, C, and I inputs used for level-sensitive scan) in the E performance level driving a load of four standard loads. Extract the maximum setup time and clock-to-Q (i.e., E-$L2$) delays from the data sheet assuming worst-case process and environment.

[20]   **1.7**   You are building the Motoroil 68W86 processor for embedded automotive applications. Suppose your system is built using flip-flops with $\Delta_{DC} = 100$ ps and $\Delta_{CQ} = 120$ ps and that there is no clock skew. Your computation requires 3 ns to complete. You are considering various pipeline organizations in which the computation is done in 1, 2, 3, or 4 clock cycles in the hopes that breaking the computation into multiple cycles would allow a faster clock rate. Make a table showing the maximum possible clock frequency for each option. Also show the total latency of the computation (in picoseconds) for each option.

[15]   **1.8**   In Exercise 1.7 we assumed that logic could be divided into any number of cycles without waste. In practice, you may have time remaining at the end of a cycle for half a gate delay; this time goes unused because it is impossible to build half a gate. Redo the exercise if on average 50 ps at the end of each cycle goes unused due to imbalanced logic.

[10]   **1.9**   Why is sequencing overhead a more important concern to designers in the year 2000 than it was in 1990?

[35]   **1.10** Gather data to extend the plots in Figures 1.3 and 1.4 from 1995 to the present. How fast is microprocessor performance increasing? How about clock rate? How long does it take for performance to double? Clock

IBM

Propagation Delays

Path (Input to Output)	Performance Level	Parameter	Delay (ns) = intercept + slope (N_{std})			
			$V_{dd} = 1.65V$ $T_j = 125°C$ Process = Slow	$V_{dd} = 1.8V$ $T_j = 25°C$ Process = Nom.	$V_{dd} = 1.8V$ $T_j = 0°C$ Process = Fast	
E-L2	E	t_{PLH}	$0.371 + 0.014N_{std}$	$0.234 + 0.009N_{std}$	$0.161 + 0.007N_{std}$	
		t_{PHL}	$0.323 + 0.009N_{std}$	$0.203 + 0.006N_{std}$	$0.139 + 0.005N_{std}$	
	H	t_{PLH}	$0.384 + 0.007N_{std}$	$0.243 + 0.005N_{std}$	$0.167 + 0.003N_{std}$	
		t_{PHL}	$0.338 + 0.004N_{std}$	$0.212 + 0.003N_{std}$	$0.146 + 0.003N_{std}$	
	J	t_{PLH}	$0.413 + 0.003N_{std}$	$0.262 + 0.002N_{std}$	$0.181 + 0.002N_{std}$	
		t_{PHL}	$0.370 + 0.002N_{std}$	$0.233 + 0.002N_{std}$	$0.161 + 0.001N_{std}$	

Capacitance (in units of N_{std}) and Cell Sizes

Input Pins	Performance Level		
	E	H	J
A	1.107	1.107	1.107
B	0.444	0.444	0.454
C	0.435	0.435	0.435
D	0.967	0.968	0.967
E	1.044	1.044	1.044
EN	0.781	0.781	0.781
I	0.393	0.393	0.393
Internal	13.153	14.285	17.036
Cell Units	18 cells	18 cells	19 cells

Latch Setup and Hold Delays (ns)

Condition[1]	Performance Level		
	E	H	J
Setup	0.0182729	0.0182729	0.0182729
Hold	0.0296109	0.0296109	0.028629

[1]. Setup and hold calculated using worst case conditions; clock and data slew rate = 0.2 ns.

IBM

Cell: LDE0001

Function: D Mimic FF, LSSD, w/Clock Enable +L2 Output

Description:

This is a combination of a clock splitter and a LPH0001 L1/L2 latch. The operation is a D mimic flip-flop. When the EN pin is low, the E clock to the latch is disabled. To avoid chopping of the clocks to the L1 and L2 portions of the latch, the state of the EN pin should only be changed while the E pin is low.

A	A clock
B	B clock
C	C clock
D	Data in
E	E clock
EN	Clock enable
I	Scan-in
L2	+L2 output (in phase with respect to input)

LSSD Latch L1 Truth Table

Inputs						L1 State
EN	A	I	C	D	E	
X	0	X	0	X	X	NC
X	1	X	0	X	X	I
X	X	0	X	X	0	D
1	0	X	1	X	1	NC
0	0	X	1	X	X	D

LSSD Latch L2 Truth Table

Inputs					Output
L1	B	C	E	EN	L2
X	0	X	X	X	NC
X	1	1	0	X	L1
X	1	X	1	1	L1
X	1	1	X	0	NC
X	1	0	X	X	L1

Figure 1.18 Excerpt from IBM SA-27 0.16-micron cell library data book (Reproduced by permission from *http://www.chips.ibm.com/techlib/products/asics/databooks.html*. Copyright 2001 by International Business Machines.)

rate? Have these trends accelerated or decelerated relative to the performance increases between 1985 and 1997? How do they compare with the SIA roadmap in Table 1.2?

[10]   **1.11** Why are domino circuits faster than static CMOS circuits?

[15]   **1.12** You are designing the Pentagram IV Processor. Consider using a pipeline built with traditional domino circuits. The pipeline requires 1 ns of logic in each cycle. Suppose the setup time of each latch is 90 ps and the clock skew is 100 ps. What is the cycle time? What fraction of the cycle is lost to sequencing overhead?

[15]   **1.13** Repeat Exercise 1.12 if the system is built from two-phase skew-tolerant domino circuits.

[20]   **1.14** You are developing a circuit methodology for an ultra-high-speed processor. You are weighing whether to recommend static CMOS circuits or traditional domino circuits. You have determined that the setup time and D-to-Q delay of your latches are approximately equal. You determined that the ALU self-bypass path is a key cycle-limiting path with a logic delay of Δ_{logic} if you construct it with static CMOS circuits. If you use domino circuits, you determine it will require only $0.7\,\Delta_{\text{logic}}$ but you must also budget clock skew. Figure 1.19 shows a design space of logic delay and clock skew. Divide the space into two regions based on whether static CMOS or traditional domino circuits offer higher operating frequencies for the given logic delay and skew. Assume there is no imbalanced logic.

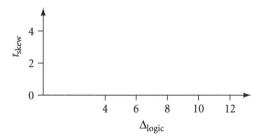

Figure 1.19   Design space of logic delay and clock skew

[15]   **1.15** Repeat Exercise 1.14 assuming that there is 0.25 FO4 delays of time wasted on average from imbalanced logic in the traditional domino design.

[10]   **1.16** Which of the circuits in Figure 1.20 produce outputs Q that can correctly drive a domino gate that evaluates on ϕ? Assume the inputs A and B to each circuit are monotonically rising while ϕ is high. The NAND and NOT gates are built from static CMOS.

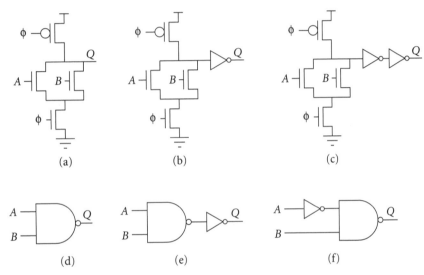

Figure 1.20   Which of these gates can properly drive a ϕ domino gate?

[15]   **1.17** Define "hard edges" as used in this chapter in your own words. Which of the following pipeline design approaches introduce hard edges: static CMOS with flip-flops, static CMOS with transparent latches, traditional domino circuits, skew-tolerant domino circuits? Why do hard edges reduce performance?

2
Static Circuits

This chapter explores the design of systems with static CMOS logic and memory elements: flip-flops, transparent latches, and pulsed latches. Most students are taught that the purpose of memory elements is to retain state. The chapter begins by arguing that memory elements can be understood more easily if viewed as a way of enforcing sequencing. It then defines some terminology regarding confusing meanings of the words "static" and "dynamic." The emphasis of the chapter is on comparing the operation of flip-flops, transparent latches, and pulsed latches. We look at how sequencing overhead, time borrowing, and hold time requirements affect each type of memory element. We then survey the literature about circuit implementations of the memory elements. The chapter concludes with a historical perspective of memory elements used in commercial processors and recommendations of what you should use for your systems.

Clocking and memory elements are highly interdependent. In general, we will assume that a single clock is distributed globally across the system and that it may be locally complemented or otherwise modified as necessary. We will look at clock generation and distribution more closely in Chapter 5.

2.1   Preliminaries

Before examining specific static circuits, let's begin with a more philosophical discussion of the purpose of memory elements and the often contradictory terminology of static and dynamic elements.

2.1.1   Purpose of Memory Elements

A static CMOS logic system consists of blocks of static CMOS combinational logic interspersed with memory elements. The purpose of the memory elements is not so much to remember anything as to enforce sequencing, distinguishing *this* from *previous* and from *next*. If memory elements were not used, fast signals might race ahead and catch up with slow signals from a different operation, resulting in hopeless confusion. Thus, an ideal memory element should slow down the early signals while adding no delay to the signals that are already late. Real memory elements generally delay even late elements; this wasted time is the sequencing overhead. We would like to build systems with the smallest possible sequencing overhead.

This idea of using memory elements for sequencing at first might seem to conflict with the idea that memory elements hold state, as undergraduates learn when studying finite state machines (FSMs). This conflict is illusory. Any static CMOS circuit can hold state: if the inputs do not change, the output will not change. The real issue is controlling when a system advances from one state to another. From this point of view, it is clear that the role of memory elements in FSMs is also to enforce sequence of state changes.

Another advantage of looking at memory elements as enforcing sequencing rather than storing state is that we can more clearly understand more unconventional circuit structures. Flip-flop systems can be equally well understood from the state storage or sequencing perspectives, partially explaining why they are so easy for designers to grasp. However, transparent latch and pulsed latch systems use two and one latch per cycle, respectively. If you attempt to point to a place in the circuit where state is stored, it is easy to become confused. Later in this chapter, we will show how both methods are adequate to enforce sequencing, which is all that really matters. Similarly, designers who believe latches are necessary to store state will be very confused by skew-tolerant domino circuits that eliminate the latches. On the other hand, it is easier to see that skew-tolerant domino circuits properly enforce sequencing and thus operate correctly.

Finally, this view of sequencing dispels some myths about asynchronous systems. Some asynchronous proponents argue that asynchronous design is good because it eliminates clocking overhead by avoiding the distribution of high-speed clocks [66]. However, when we view memory elements as sequencing devices, we see that all systems, synchronous or asynchronous, must pay some sequencing overhead because it is impossible to slow the early signals without slightly delaying the late signals. Asynchronous designs merely replace the problem of distributing high-speed clocks with the problem of distributing high-speed sequencing control signals.

2.1.2 Terminology

We will frequently use the terms "static" and "dynamic," which unfortunately have two orthogonal meanings. Most of the time we use the terms to describe the type of circuit family. *Static circuits* refer to circuits in which the logic gates are unclocked: static CMOS, pseudo-NMOS, pass-transistor logic, and so on. *Dynamic circuits* refer to circuits with clocked logic gates: especially domino, but also Zipper logic and precharged RAMs and programmable logic arrays (PLAs). A second meaning of the words is whether a memory element will retain its value indefinitely. *Static storage* employs some sort of feedback to retain the output indefinitely even if the clock is stopped. *Dynamic storage* stores the output voltage as charge on a capacitor, which may leak away if not periodically refreshed.

To make matters more confusing, a particular circuit may independently be described as using static or dynamic logic and static or dynamic storage. For example, most systems are built from static logic with static storage. However, the Alpha 21164 uses blocks of static logic with dynamic storage [26], meaning that the processor has a minimum operating frequency so that it does not lose data stored in the latches. The Alpha 21164 also uses blocks of domino (i.e., dynamic logic) with dynamic storage. Most other processors use domino with static storage so that if they were placed in a notebook computer and the clock was stopped during sleep mode, the processor could wake up and pick up from where it left off without having lost information.

In this book, the terms "static" and "dynamic" will generally be used in the first context of circuit families. This chapter describes the sequencing of static circuits. The next chapter describes the sequencing of domino (dynamic) circuits. In each chapter, we will explore the transistor implementations of storage elements, in which the second context of the words is important. Generally we will present the dynamic form of each element first, then discuss how feedback can be introduced to staticize the element.

2.2   Static Memory Elements

Although there are a multitude of possible memory elements, including S-R latches and J-K flip-flops, most CMOS systems are built with just three types of memory elements: edge-triggered flip-flops, transparent latches, and pulsed latches. Transparent and pulsed latch systems are sometimes

called two- and one-phase latch systems, respectively. There is some confusion about terminology in the industry, which this section seeks to clear
up; in particular, pulsed latches are commonly and confusingly referred
to as "edge-triggered flip-flops." All three elements have clock and data
inputs and an output. Depending on the design, the output may use true,
complementary, or both polarities.

The section begins with timing diagrams illustrating the three memory elements. It then analyzes the sequencing overhead of each element.
Latches are particularly interesting because they allow time borrowing,
which is described in more detail. Finally, the min-delay issues involving
hold time are described.

2.2.1   Timing Diagrams

Edge-triggered flip-flops are also known as master-slave or D flip-flops.
Their timing is shown in Figure 2.1. When the clock rises, the data input
is sampled and transferred to the output after a delay of Δ_{CQ}. At all other
times, the data input and output are unrelated. For the correct value to be
sampled, data inputs must stabilize a setup time Δ_{DC} before the rising
edge of the clock and must remain stable for a hold time Δ_{CD} after the rising edge.

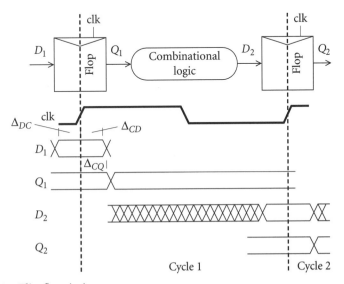

Figure 2.1   Flip-flop timing

We can check that memory elements properly sequence data by making sure that data from *this* cycle does not mix with data from the *previous* or *next* cycles. This is clear for a flip-flop because all data advances on the rising edge of the clock and at no other time.

Transparent latches are also known as half-latches, D-latches, or two-phase latches. Their timing is shown in Figure 2.2. When the clock is high, the output tracks the input with a lag of Δ_{DQ}. When the clock is low, the output holds its last value and ceases to track the input. For the correct value to be sampled, the data must stabilize a setup time Δ_{DC} before the falling edge of the clock and must remain stable for a hold time Δ_{CD} after the falling edge. It is confusing to call a latch ON or OFF because it is not clear whether ON means that the latch is passing data or ON means that the latch is latching (i.e., holding old data). Instead, we will use the terms "transparent" and "opaque." To build a sequential system, two half-latches must be used in each cycle. One is transparent for the first part of the cycle, while the other is transparent for the second part of the cycle using a locally complemented version of the clock, clk_b.

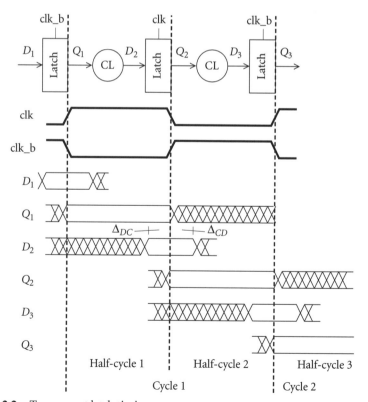

Figure 2.2   Transparent latch timing

Latches may be placed at any point in the half-cycle; the only constraint is that there must be one latch in each half-cycle. Many designers think of latches at the end of the half-cycle. In Section 1.3, we illustrated latches in the middle of the half-cycle. We will discuss the relative advantages of these choices in Section 4.1. In any event, sequencing is enforced because there is always an opaque latch between data in one cycle and data in another cycle.

Pulsed latches, also known as one-phase or glitch latches, behave exactly as transparent latches, but are given a short pulse of width t_{pw} produced by some local clock generator instead of the usual 50% duty cycle clock. Therefore, only one pulsed latch is necessary in each cycle. Their timing is shown in Figure 2.3. Data must set up and hold around the falling edge of the pulse, just as it does around the falling edge of a transparent latch clock.

Pulsed latches enforce sequencing in much the same way as flip-flops. If the pulse is narrow enough, it begins to resemble an edge trigger. For this reason, pulsed latches are sometimes misleadingly referred to as "edge-triggered flip-flops." We will be careful to avoid this because the sequencing overhead differs in important ways. As long as the pulse is shorter than the time for data to propagate through the logic between pulsed latches, the pulse will end before new data arrives and data will be safely sequenced.

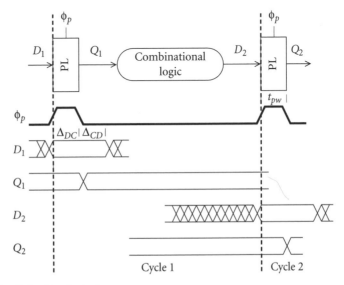

Figure 2.3   Pulsed latch timing

2.2.2   Sequencing Overhead

Ideally, the cycle time of a system should equal the propagation delay through the longest logic path. Unfortunately, real sequential systems introduce overhead that increases the cycle time from three sources: propagation delay, setup time, and clock skew. In this section, we will look at the sequencing overhead of each type of memory element, obtaining equations in the form

$$T_c = \Delta_{\text{logic}} + \text{overhead} \tag{2.1}$$

In Section 1.1 we found that flip-flops pay all three sources of overhead. Data is launched on the rising edge of one flip-flop. It must propagate to the Q output, then pass through the logic, and then arrive at the next flip-flop a setup time before the clock rises again. Any skew between the clocks that could cause the first flip-flop to fire late or the second flip-flop to sample early must be budgeted in the cycle time T_c:

$$T_c = \Delta_{\text{logic}} + \Delta_{CQ} + \Delta_{DC} + t_{\text{skew}} \tag{2.2}$$

In Section 1.3 we saw that transparent latches can be used such that data arrives more than a setup time before the falling edge of the clock, even in the event of clock skew. The data will promptly propagate through the latch and subsequent logic may begin. Therefore, neither setup time nor clock skew must be budgeted in the cycle time. However, there are two latches in each cycle, so two latch propagation delays must be included:

$$T_c = \Delta_{\text{logic}} + 2\Delta_{DQ} \tag{2.3}$$

There is widespread misunderstanding about the overhead in transparent latch systems. Many designers and even textbooks [92] mistakenly budget setup time or clock skew in the cycle time. Worse yet, some realize their error but continue to budget setup time on the grounds that they are "being conservative." In actuality, they are just forcing overdesign of the sections of the chip that use transparent latches. This overdesign leads to more area, higher power, and longer time to market. They may argue that the chip will run faster if the margin was unnecessary, but if the chip contains any critical paths involving domino or flip-flops, those paths will limit cycle time and the margin in the transparent latch blocks will go unused. It is much better to add margin only for overhead that actually exists.

Also, notice that our analysis of latches did not depend on the duty cycles of the clocks. For example, the latches could be controlled by non-overlapping clocks without increasing the cycle time. Nonoverlapping clocks will be discussed further in Section 2.2.4 and in Chapter 6.

Pulsed latches are the most interesting to analyze because their overhead depends on the width of the pulse. Data must set up before the falling edge of the pulse, even in the event of clock skew. This time required may be before or after the actual rising edge of the pulse, depending on pulse width and clock skew, as illustrated in Figure 2.4. If the pulse is wider than the setup time plus clock skew, the data can arrive while the pulsed latch is transparent and pass through with only a single latch propagation delay. When the pulse is narrow, data must arrive by a setup time and clock skew before the nominal falling edge. This required time can be earlier than the actual rising edge of the pulse. Therefore, the data may sit idle for some time until the latch becomes transparent, increasing the overhead in the case of short pulse widths. Later, we will see that long pulse widths cause hold time difficulties, so pulsed latches face an inherent trade-off:

$$T_c = \Delta_{\text{logic}} + \Delta_{DQ} + \max(0, \Delta_{DC} + t_{\text{skew}} - t_{pw}) \tag{2.4}$$

In summary, flip-flops always have the greatest sequencing overhead and are thus costly for systems with short cycle times in which sequencing overhead matters. Nevertheless, they remain popular for lower-performance systems because they are well understood by designers,

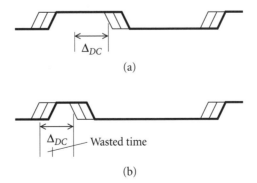

Figure 2.4   Effect of pulse width on sequencing overhead: pulse width greater than setup time (a) and pulse width less than setup time (b)

require only a single clock wire, and are supported by even the least-sophisticated timing analyzers. Transparent latches require two latch propagation delays, but hide clock skew from the cycle time. Pulsed latches are potentially the fastest because with a sufficiently wide pulse they can hide clock skew and only introduce a single latch propagation delay. This speed comes at the expense of strict min-delay constraints.

2.2.3 Time Borrowing

We have seen that a principal advantage of transparent latches over flip-flops is the softer edges that allow data to propagate through the latch as soon as it arrives instead of waiting for a clock edge. Therefore, logic does not have to be divided exactly into half-cycles. Some logic blocks can be longer while others are shorter, and the latch-based system will tend to operate at the average of the delays; a flip-flop–based system would operate at the longest delay. This ability of slow logic in one half-cycle to use time nominally allocated to faster logic in another half-cycle is called *time borrowing* or *cycle stealing*.

The exact use of time borrowing depends on the placement of latches. If latches are nominally placed at the beginning of half-cycles, as was done in Figure 2.2, we say that when data arrives late from one half-cycle, it can borrow time forward into the next half-cycle. If latches are nominally placed at the end of half-cycles, we say that if the data arrives early at the latch, the next half-cycle can borrow time backward by starting as soon as the data arrived instead of waiting for the early half-cycle to end. In systems constructed entirely of transparent latches, it does not matter whether we think of time borrowing as forward or backward. In systems interfacing transparent latches to flip-flops or domino gates that introduce hard edges, the direction of time borrowing becomes more important, as we shall see in Section 4.1.1.

Moreover, time borrowing may operate over multiple half-cycles. In the case of forward borrowing, data may arrive late at a latch. If the logic after the latch requires exactly half of a cycle, the result will arrive late at the following latch. This borrowing may continue indefinitely so long as the data never arrives so late at a latch that the setup time is violated.

Let us examine how much time borrowing is possible in any half-cycle for transparent latches. For the purpose of illustration, assume that data departs a latch in half-cycle 1 at the rising edge of the clock, as shown in

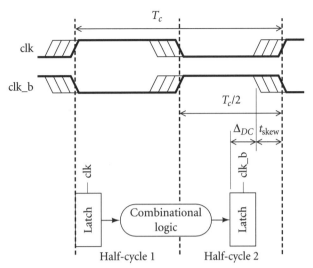

Figure 2.5 Calculation of maximum time borrowing for transparent latches

Figure 2.5. The data nominally arrives at the next latch exactly $T_c/2$ later. However, the circuit will operate correctly so long as the data arrives before the next latch becomes opaque. The difference between the actual arrival and the nominal arrival is the amount of time borrowing. The latest possible arrival time must permit the latch to set up before the earliest skewed receiver clock. Therefore, the maximum amount of time that can be borrowed is

$$t_{\text{borrow}} = \frac{T_c}{2} - t_{\text{skew}} - \Delta_{DC} \tag{2.5}$$

In the limit of long cycle time T_c, this approaches half a cycle. For shorter cycles, the clock skew and setup time overhead reduce the amount of available time borrowing.

   If the latches are transparent for shorter amounts of time, the amount of time available for borrowing is reduced correspondingly. For example, in the next section we will discuss the possibility of using nonoverlapping clocks to avoid min-delay problems. The nonoverlap is subtracted from the amount of time available for borrowing. Similarly, pulsed latches can be viewed as transparent latches with short transparency windows. The amount of time borrowing supported by a transparent latch is therefore

$$t_{\text{borrow}} = t_{pw} - t_{\text{skew}} - \Delta_{DC} \tag{2.6}$$

If this amount is negative, the pulse width is too narrow to allow time borrowing. Of course, flip-flops permit no time borrowing because they impose hard edges.

Equations 2.5 and 2.6 show that there is a direct trade-off between time borrowing and clock skew. In effect, skew causes uncertainty in the arrival time of data. Budgeting for this uncertainty is just like budgeting for data intentionally being late on account of time borrowing. By more tightly bounding the amount of clock skew in a system, more time borrowing is possible.

Time borrowing may be used in several ways. Designers may *intentionally* borrow time to balance logic between slower and faster pipeline stages. The circuits will *opportunistically* borrow time after fabrication when manufacturing and environmental variations and inaccuracies in the analysis tools cause some delays to be faster and other delays to be shorter than the designer had predicted. While in principle designers could always repartition logic to balance delay as accurately as they can predict in advance and avoid intentional time borrowing, such repartitioning takes effort and increases time to market. Moreover, opportunistic time borrowing is always a benefit because it averages out uncertainties that are beyond the control of the designer and that would have otherwise limited cycle time.

A practical problem with time borrowing is that engineers faced with critical paths may assume they will be able to borrow time from an adjacent pipeline stage. Unfortunately, the engineer responsible for the adjacent stage may also have assumed that she could borrow time. When the pieces are put together and submitted for timing analysis, the circuit will be very far from meeting timing. This could be solved by good communication between designers, but because of trends toward enormous design teams and because of the poor documentation about such assumptions and the turnover among engineers, mistakes are often made. Design managers who have been burned by this problem in the past tend to forbid the use of time borrowing until very late in the design when all other solutions have failed. This is not to say that time borrowing is a bad thing; it simply must be used wisely.

2.2.4   Min-Delay

So far, we have focused on the question of max-delay: how long the cycle must be for each memory element to meet its setup time. The max-delay

constraints set the performance of the system, but are relatively innocu-
ous because if they are violated, the circuit can still be made to function
correctly by reducing the clock frequency. In contrast, circuits also have
min-delay constraints that memory element inputs must not change until
a hold time after the sampling edge. If these constraints are violated, the
circuit may sample the output while it is changing, leading to incorrect
results. Min-delay violations are especially insidious because they cannot
be fixed by changing the clock frequency. Therefore, the designer is forced
to be conservative.

Figure 2.6 shows how min-delay problems can lead to incorrect operation of flip-flops. In the example, there are two back-to-back flip-flops
with no logic between them. This is common in pipelined circuits where
information such as an instruction opcode is carried from one pipeline
stage to the next without modification as the instruction is processed.
Suppose data input D_1 is valid for a setup and hold time around the rising
edge of clk, but that the propagation delay to Q_1 is particularly short. Q_1 is
the input to the second flip-flop and changes before the end of the hold
time for the second flip-flop. Therefore, the second flip-flop may incorrectly sample this new data and pass it on to Q_2. In summary, the data
that was at input D_1 before the clock edge arrives at not only Q_1 but also
Q_2 after the clock edge. This is referred to as double-clocking, a hold time
or min-delay violation, or a race.

The term "min-delay" comes from the fact that the problem can be
avoided by guaranteeing a minimum amount of delay between consecutive flip-flops. If there were more delay between the rising edge of the

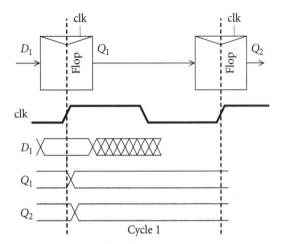

Figure 2.6   Min-delay problems in flip-flops

clock and the time data arrived at the second flip-flop, the hold time would not have been violated and the circuit would have worked correctly.

Min-delay problems are exacerbated by clock skew. If skew causes the clock of the first flip-flop to rise early, its output will become valid early. If skew then also causes the clock of the second flip-flop to rise late, its input will have to hold until a later time. Therefore, more delay is necessary between the flip-flops to ensure the hold time is not violated. Clock skew can be viewed as increasing the effective hold time of the second memory element.

We can guarantee that min-delay problems will never occur by checking a simple delay constraint between each pair of consecutive memory elements. Assume that data departs the first element as early as possible. Add the shortest possible delay between this departure time and the arrival at the second element; this is called the *contamination delay*. The arrival must be at least a hold time after the sampling edge of the second element, assuming maximum skew between the elements. To analyze our prospective latching techniques, we need a few more definitions. Let us define δ_{CQ} as the contamination delay of the memory element, that is, the minimum time from the clock switching until the output becoming valid. This is like Δ_{CQ} but represents the minimum instead of maximum delay. Let δ_{logic} be the contamination delay through the logic between the memory elements.

For flip-flops, data departs the first flip-flop on the rising edge of the clock. The flop and logic contamination delays must be adequate for the data to arrive at the second flip-flop after its hold time has elapsed, even budgeting clock skew:

$$\delta_{CQ} + \delta_{logic} \geq \Delta_{CD} + t_{skew} \tag{2.7}$$

Solving for the minimum logic contamination delay, we find

$$\delta_{logic} \geq \Delta_{CD} + t_{skew} - \delta_{CQ} \tag{2.8}$$

Notice that the constraint is independent of cycle time T_c. As expected, min-delay problems cannot be fixed by adjusting the cycle time.

For latches, data departs the first latch on the rising edge of one half-cycle. The latch and logic contamination delays must be sufficient for the data to arrive a hold time after the falling edge of the previous half-cycle.

Let us define $t_{\text{nonoverlap}}$ as the time from the falling edge of one half-cycle to the rising edge of the next. This time is typically 0 for complementary clocks, but may be positive for nonoverlapping clocks. The minimum logic contamination delay is

$$\delta_{\text{logic}} \geq \Delta_{CD} + t_{\text{skew}} - \delta_{CQ} - t_{\text{nonoverlap}} \tag{2.9}$$

Notice that this minimum delay is through each half-cycle of logic. Therefore the full cycle requires minimum delay twice as great.

The following example may clarify the use of nonoverlapping clocks:

EXAMPLE 2.1   What is the logic contamination delay required in a system using transparent latches if the hold time is 0, the latch contamination delay is 0.5 FO4 inverter delays, the clock skew is 1 FO4 delay, and the nonoverlap is 2 FO4 delays, as shown in Figure 2.7?

SOLUTION δ_{logic} must be at least $0 + 1 - 0.5 - 2 = -1.5$ FO4 delays. Because logic delays are always nonnegative, it is impossible for this system to experience min-delay problems.   ∎

Two-phase nonoverlapping clocking was once popular because of min-delay safety. It is still a good choice for student projects because it is completely safe; by using external control of the clock waveforms, the student can always provide enough nonoverlap and slow-enough clocks to avoid problems with either min-delay or max-delay. However, commercial high-speed designs seldom use nonoverlapping clocks because it is easier to distribute a single clock globally, and then locally invert it to obtain the two latch phases. Instead, the commercial designs check min-delay and insert buffers to increase delay in fast paths. Nonoverlapping clocks also reduce the possible amount of time borrowing. Note that there is a common fallacy that nonoverlapping clocks allow less time for useful computation. As can be seen from Figure 2.7, this is not the case; the full cycle less two latch delays is still available. The only penalty is the reduced opportunity for time borrowing.

For pulsed latches, data departs the first latch on the rising edge of the pulse. It must not arrive at the second pulsed latch until a hold time after the falling edge of the pulse. As usual, the presence of clock skews between the pulses increases the hold time. Therefore, the minimum contamination delay is

$$\delta_{\text{logic}} \geq t_{pw} + \Delta_{CD} + t_{\text{skew}} - \delta_{CQ} \tag{2.10}$$

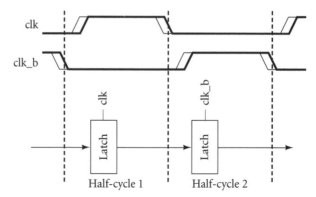

Figure 2.7   Transparent latches with nonoverlapping clocks

This is the largest required contamination delay of any latching scheme. It shows the trade-off that although wider pulses can hide more clock skew and even permit small amounts of time borrowing, the wide pulses increase the minimum amount of delay between latches. Adding this amount of delay between pulsed latches in cycles that perform no logic can take a significant amount of area. Therefore, systems that use pulsed latches for the critical paths that require low sequencing overhead sometimes also use flip-flops to reduce min-delay problems on paths that merely stage data along without processing.

You may have noticed that flip-flops and pulsed latches have a minimum delay per cycle, while transparent latches have a minimum delay per half-cycle, and hence about twice as much minimum delay per cycle. This may seem strange because flip-flops can be built from a pair of back-to-back transparent latches. Why should flip-flops have half the min-delay requirement as transparent latches if the systems have exactly the same building blocks? The answer is that flip-flops are usually constructed with zero skew between adjacent latches. By making the hold time Δ_{CD} less than the contamination delay δ_{DQ}, the minimum logic delay between the two latches in the flip-flop is negative. If this were not the case, flip-flops would insidiously fail by sampling the input on the falling edge of the clock as well as the rising edge! We will revisit this issue while discussing flip-flop design in Section 2.3.3.

Min-delay can be enforced in many short paths by adding buffers. Long channel lengths are often used to make slower buffers so that fewer buffers are required. The hardest min-delay problems occur in paths that could be either fast or slow in a data-dependent fashion. For example, a path built from a series of NAND gates may be fast when both parallel PMOS transistors turn on and slower when only one PMOS transistor turns on. A path using wide domino OR gates is even more sensitive to input patterns. Therefore, circuit designers occasionally encounter paths that have both min- and max-delay problems. Because buffers cannot be added without exacerbating the max-delay problem, the circuits may have to be redesigned.

Min-delay requirements are easy to check because they only involve delays between pairs of consecutive memory elements. They are also conservative for systems that permit time borrowing because they assume data always departs the first latch at the earliest possible time. In a real system, time borrowing may cause data to depart the first latch somewhat later, making min-delay easier to satisfy. Unfortunately, if the real system is operated at reduced frequency, or at higher voltage where transistors are faster, data may again depart the first latch at the earliest possible time. Therefore, it is unwise to depend on data departing late to guarantee min-delay.

Because min-delay violations result in nonfunctional circuits at any operating frequency, it is necessary to be conservative when checking and guaranteeing hold times. Discovering min-delay problems after receiving chips back from fabrication is extremely expensive because the violation must be fixed and new chips must be built before the debugging of other problems such as long paths or logic errors can begin. This may add two to four months to the debug schedule in an industry with product cycles of two years or less.

2.3 Memory Element Design

Now that we have discussed the performance of various memory elements, we will turn to the transistor-level implementations. Many transparent latch implementations have been proposed, but in practice a very old, simple design is generally best. Pulsed latches can be built from transparent latches with a brief clock pulse or may integrate pulsing circuitry into the latch. Flip-flops can be composed of back-to-back latches or of various precharged structures. We will begin by showing dynamic versions of the memory elements, and then discuss how to staticize the latches.

2.3.1   Transparent Latches

The simplest dynamic latch dating back to the days of NMOS is just a pass transistor, shown in Figure 2.8. Its output can only swing from 0 to $V_{DD} - V_t$, where V_t is the threshold voltage of the transistor. To provide rail-to-rail output swings, a full transmission gate is usually a better choice.

Such latches have many drawbacks. The latch must drive only a capacitive load. If it drove the diffusion input of another pass transistor, when the latch is opaque and the pass transistor turns on, charge stored on the output node of the latch is shared between the capacitances on both sides of the pass transistor, causing a voltage droop. This effect is called *charge sharing* and can corrupt the result. Coupling onto the output node from adjacent wires is also a problem because the output is a dynamic node. Latch setup time depends on both the load and driving gate, just as the delay through an ordinary transmission gate depends on the driver and load. Therefore the latch is hard to use in a standard cell methodology that defines gate delays without reference to surrounding gates.

The setup time issue can be solved by placing an inverter on the input. Sometimes the inputs to the transmission gate are separated to create a tristate inverter instead, as shown in Figure 2.9. The performance of both designs is comparable; removing the shorting connection slightly reduces drive capability, but also reduces internal diffusion parasitics. Note that it is important for the clock input to be on the inside of the stack. If the data input were on the inside, changes in the data value while the latch was off could cause charge sharing and glitches on the output.

Placing an inverter on the output solves the charge-sharing problems and reduces noise coupling because the dynamic node is not routed over long distances.

To minimize clock skew, most chips distribute only a single clock signal across the entire chip. Locally, additional phases are derived as neces-

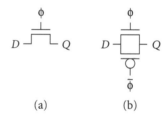

(a)   (b)

Figure 2.8   Simple latches: pass-transistor latch (a) and transmission gate latch (b)

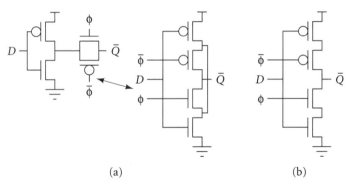

Figure 2.9   Latches built as inverter + transmission gate (a) versus tristate (b)

sary. For example, an inverter may be used to create clk_b from clk. Once upon a time, some designers had only considered globally distributing the multiple phase. They correctly argued that such global distribution would lead to severe clock skew between the phases and high sequencing overhead [96]. Therefore, they recommended using as few clock phases as possible. This reasoning is faulty because local phase generators can provide many clock phases with relatively low skew. Nevertheless, it led to great interest in latching with only a single clock phase. Such true single-phase clocking (TSPC) latches are shown in Figure 2.10 as they compare to traditional latches with local phase generators.

TSPC latches have a longer propagation delay because the data passes through more gates. As designers realized that inverting the clock can be

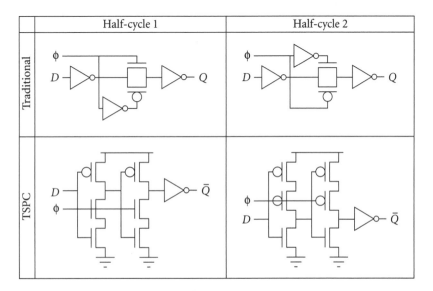

Figure 2.10   TSPC versus traditional latches

done relatively cheaply and that TSPC latches are slower and larger than normal latches, TSPC latches fell from commercial favor. For example, the DEC Alpha 21064 used TSPC latches. A careful study after the project found that TSPC consumed 25% more area, presented 40% more clock load, and was 20% slower than traditional latches. Therefore, the Alpha 21164 returned to traditional latch design [26, 28].

All latches that have been presented so far are dynamic; if left indefinitely, subthreshold conduction through the nominally OFF transistors may cause the gate to leak to an incorrect value. To allow a system to run at low frequency or with a stopped clock, the dynamic node must be staticized. This can be done with a weak feedback inverter or with a tristate feedback inverter. Weak inverter "jamb" latches introduce a ratio problem: the transistors in the forward path must be strong enough to overpower the feedback inverter over all values of PMOS-to-NMOS mobility ratios. Tristate inverters are thus safer and avoid contention power, but are larger. Feeding back from the output inverter is risky because noise on the output can overwrite the contents of the latch via the feedback gate while the latch is opaque. Thus, the output is often buffered with a separate gate, as shown in Figure 2.11. Staticizing TSPC latches is more cumbersome because there are multiple dynamic nodes.

The setup time Δ_{DC} of the latch is the time required for data to propagate from the input to the dynamic node after the transmission gate so that when the clock falls, the new data sits on the dynamic node. The propagation delay Δ_{DQ} is the time required for data to propagate from the input all the way to the output. The hold time Δ_{CD} is zero or even negative in many latch implementations like those of Figure 2.11 because if data arrives at the input at the same time the clock falls, the transmission gate will be OFF by the time the data propagates through the input inverter.

In summary, for highest-performance custom design, a bare transmission gate is very fast and was used by DEC in the Alpha 21164. Care should be taken to compute the setup time as a function of the driver as well as the load capacitance. For synthesized design, an inverter/transmission gate/inverter combination latch is safer to guarantee setup times and safety of the dynamic node. In any case, designs that must support stop clock or low-frequency operation should staticize the dynamic node with a feedback gate.

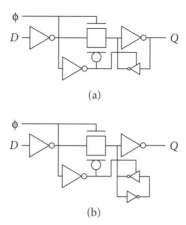

(a)

(b)

Figure 2.11   Staticizing latches: noise on Q fed back to latch node (a) and harmless noise on Q (b)

2.3.2   Pulsed Latches

In principle, a pulsed latch is identical to a transparent latch, but receives a narrow clock pulse instead of a 50% duty cycle clock [89]. The key challenge to pulsed latches is generating and distributing a pulse that is wide enough for correct operation but not so wide that it causes terrible mindelay problems. Unfortunately, pulse generators are subject to variation in transistor parameters and operating conditions that may narrow or widen the pulse. If you do not remember that setup times will also get shorter as transistors become faster and pulses narrow, you may conclude it is virtually impossible to design a reasonable pulse generator. Commercial processors have proved that the concept can work, though we will see in the historical perspective that these processors have had production problems that might be attributed to risky circuit techniques. Because it is extremely difficult to propagate narrow pulses cleanly across long lossy wires, we will assume the pulses must be generated locally.

The conceptually simplest pulsed latch is shown in Figure 2.12. The left portion is called a clock chopper, pulse generator, or one-shot; it produces a short pulse ϕ_p on the rising clock edge. The clock chopper can serve a single latch or may locally produce clocks for a small bank of latches. The latch is shown in dynamic form without an output inverter, but can be modified to be static and have better noise immunity just like a

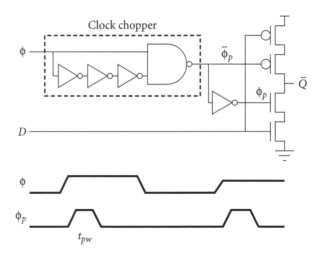

Figure 2.12   Simple pulsed latch

transparent latch. Indeed, any transparent latch, including unbuffered transmission gates and TSPC latches, may be pulsed.

It is difficult to generate very narrow pulses with a clock chopper because of the finite rise and fall times, so shorter pulses are often constructed from the intersection of two wider pulses. The Partovi pulsed latch [64] of Figure 2.13 integrates the pulse generator into the latch to achieve narrower pulses in just this way at the expense of greater Δ_{DQ} propagation delay. The latch was originally called a "hybrid latch-flip-flop" (HLFF) because it was intended to replace a flip-flop. However, calling it a flip-flop is misleading because it is not strictly edge-triggered.

When the clock is low, the delayed clock $\overline{\phi}_D$ is initially high and node X is set high. The output Q floats at its old value. When the clock rises, both ϕ and $\overline{\phi}_D$ are briefly high before $\overline{\phi}_D$ falls. During this overlap, the latch is transparent. The NAND gate acts as an inverter, passing $X = \overline{D}$ because the other two inputs are high. The final gate also acts as an inverter because transistors N_1 and N_3 are high. Therefore, the latch acts as a buffer while transparent. When $\overline{\phi}_D$ falls, the pulldown stacks of the NAND gate and the final gate both turn off. This cuts output node Q off from the input D, leaving the latch in an opaque mode. As usual, the latch can be made static by placing cross-coupled inverters on the output.

We will return to pulsed latch variants in Section 4.1.2, when we study the static-to-domino interface.

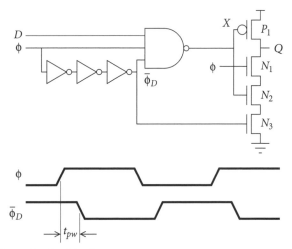

Figure 2.13 Partovi pulsed latch

2.3.3 Flip-Flops

A flip-flop can be constructed from two back-to-back transparent latches, as shown in Figure 2.14. When the clock is low, the first latch is transparent and the second latch is opaque. Therefore, data will advance to the internal node X. When the clock rises, the first latch will become opaque, blocking new inputs, and the second latch will become transparent. The flip-flop setup time is the setup time of the first latch. The clock-to-Q delay is the time from when data is at the dynamic node of the first latch and the clock rises until the data reaches the output of the flip-flop. It is therefore apparent that the sum of the setup and clock-to-Q delays of the flip-flop is equal to the sum of the propagation delays through the latches because in both cases the data must pass through two latches. Combining this observation with Equations 2.2 and 2.3, we see that the overhead of a

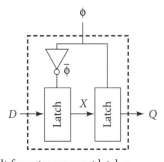

Figure 2.14 Flip-flop built from transparent latches

flip-flop system is worse than that of a transparent latch system by the clock skew.

In practice, the latches used in flip-flops can be slightly simpler than those used in stand-alone applications because the internal node X is protected and does not need all the buffering of two connected latches. Figure 2.15 shows such optimized flip-flops built from transmission gate latches and from TSPC latches.

Remember that the skew between back-to-back latches of a flip-flop must be small or the flip-flop may have an internal min-delay problem. This problem is illustrated in Figure 2.16. Suppose that $\bar{\phi}$ is badly skewed relative to ϕ, possibly because the local inverter is undersized and thus too slow. When the clock falls, both transistors P_1 and P_2 of Figure 2.15 will be simultaneously on for a brief period of time. This allows data to pass from D to $\bar{Q}$ during this time, effectively sampling the input on the falling edge of the clock. The problem can be avoided by ensuring that the $\bar{\phi}$ inverter is fast enough to turn off P_2 before new data arrives. TSPC latches are immune to this problem because they only use one clock, but are susceptible to internal races when the clock slope is very slow, causing both NMOS and PMOS clocked transistors to be on simultaneously during the transition. A modified traditional flip-flop design based on tristate latches instead of transmission gate latches shown in Figure 2.17 [24] also avoids

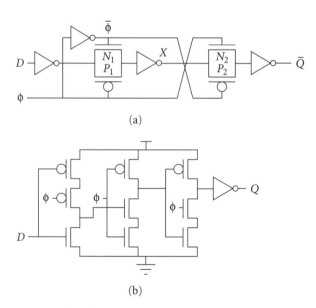

(a)

(b)

Figure 2.15   Optimized flip-flop implementations: traditional (a) and TSPC (b)

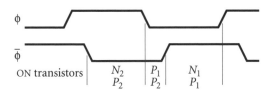

ON transistors N_2 P_1 N_1
                  P_2 P_2 P_1

Figure 2.16   Clock skew may cause internal race in flip-flops

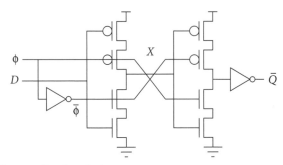

Figure 2.17   No-race flip-flop design

internal races because data will pass through the NMOS transistors of one tristate and the PMOS transistors of the other tristate, never through the PMOS transistors of both stages. Of course, while avoiding internal races is necessary, it does not eliminate the problem of min-delay between flip-flops.

The traditional flip-flop can be made static by adding feedback onto the dynamic nodes after each of the two transmission gates. This would be very costly in the TSPC flip-flop for three reasons: (1) the presence of three dynamic nodes instead of just two, (2) the lack of an inverted version of each node to feed back, and (3) the lack of a complementary clock to operate a transmission gate.

The Klass semidynamic flip-flop (SDFF) [47, 48] of Figure 2.18 is based on a different idea. Like the Partovi pulsed latch, it operates on the principle of intersecting pulses. Compared to the Partovi latch, it may have a slightly shorter propagation delay, but is edge-triggered and thus loses the skew tolerance and time-borrowing capabilities of pulsed latches. The Klass SDFF replaces the static NAND gate of the Partovi pulsed latch of Figure 2.13 with a dynamic NAND. Because node X is guaranteed to be monotonically falling while the clock is high, the output stage can also be simplified by removing N_3. Another modification is that $\bar{\phi}_D$ is gated by X. If D is low, $\bar{\phi}_D$ will fall three gate delays after the clock

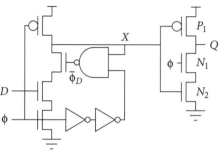

Figure 2.18 Klass semidynamic flip-flop

rises, providing a very narrow pulse. If D is high, X will start to pull down and $\bar{\phi}_D$ will not fall. This allows more time for X to fall all the way and purportedly permits a narrower pulse than would be possible if X had to pull from high all the way low during the pulse. Another advantage is that fast, relatively complex logic can be built into the first stage, which behaves as a dynamic gate. The latch needs cross-coupled inverters on both X and Q for fully static operation. A drawback relative to ordinary flip-flops is that, like a pulsed latch, the hold time is increased by the pulse width.

Yet another flip-flop design is the sense-amplifier flip-flop (SAFF) [28, 55, 58] of Figure 2.19, which has been used in the Alpha 21264 and in the StrongARM. The flip-flop requires differential inputs and produces a differential output. It can be understood as a dual-rail domino buffer with regenerative feedback followed by an SR latch on the output to retain the output state during precharge. Remarkably, a single transistor N_4 serves to staticize the latch; this transistor can be omitted in dynamic implementations.

When the clock is low, evaluation transistor N_1 is off and precharge transistors P_3 and P_4 pull the internal nodes X and $\bar{X}$ high. When the clock rises, one of the inputs will be at a higher voltage than another. This will cause the corresponding node X or $\bar{X}$ to pull down. Transistors P_1, P_2, N_5, and N_6 together form a cross-coupled inverter pair that performs regenerative feedback to amplify the difference between X and $\bar{X}$. Initially both N_5 and N_6 are on, allowing either side to pull low. As one side pulls down, the NMOS transistor on the other side begins to turn off and the PMOS transistor begins to turn on, holding the other side high. Once one side has fully pulled down, the flip-flop ceases to respond to input changes so the hold time is quite short. If the input changes, the internal

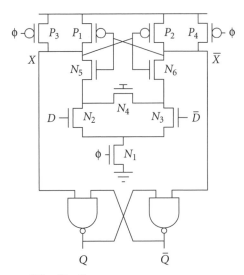

Figure 2.19   Sense-amplifier flip-flop

nodes may be left floating unless weak staticizer N_4 is available to provide
a trickle of current. When the clock falls, the internal nodes precharge but
the cross-coupled NAND gates on the output serve as an SR latch to retain
the value.

As a general-purpose flip-flop, the SAFF is not very fast. One of the
internal nodes must first pull down, causing one of the outputs to rise,
then the other output to fall, leading to three gate delays through the
flop. However, the SAFF has other advantages. It is used in the Alpha
21264 to amplify 200 mV signal swings [22] from the register file and on
other heavily loaded internal busses, greatly reducing the delay of the
input swing. Because the core of the flop is just a dual-rail domino gate,
it is easy to build logic into the gate for greater speed. Care must be
taken, however, when incorporating logic to avoid charge-sharing noise
that incorrectly trips the sense amplifier. Finally, when the flip-flop interfaces to domino logic, the SR latch can be removed because the domino
logic does not need inputs to remain stable all cycle. In summary, the
SAFF is a good choice for certain applications where its unique features
are beneficial.

Stojanovic and Oklobdzija made a thorough study of flip-flop variants
[81]. The study focused on the power-delay product rather than the delay
given fixed input and output capacitance specifications. It found the Klass
SDFF to be the fastest, while the traditional flip-flop built from two transparent latches offered the lowest power-delay product.

2.4   Historical Perspective

The microprocessor industry has experienced a fascinating evolution of latching strategies. Digital Equipment Corporation (now part of Compaq) took the microprocessor industry by surprise by developing the 200 MHz Alpha 21064 in 1992 [16], when most other microprocessors were well below 100 MHz. At that time, designers strove to distribute the clock on a single wire to nearly all latches and to avoid completely the need for checking min-delay problems with inadequate tools. They therefore challenged the conventional wisdom by using TSPC latches instead of traditional latches or flip-flops. As we have seen, TSPC proved a poor choice, requiring more area and clock power and being slower than regular transparent latches. Therefore, the Alpha 21164 returned to static latches [26]. They gained speed by using a bare transmission gate as the basic latch and choosing from a characterized library of simple gates at the input and output of the transmission gate. This allowed designers to build logic into the latch and minimize the sequencing overhead. At least one logic gate was required between each latch to avoid min-delay problems. Curiously, the Alpha 21264 began using a family of edge-triggered flip-flops [28]; overhead was tolerable because great effort went into the clock distribution system.

Interestingly, the 21064 and 21164 employed dynamic latches to avoid the extra delay of staticizers. To retain dynamic state, the processors had a minimum clocking frequency requirement as high as 1/10 of the maximum frequency [26]. This made testing and debug more challenging and is not a viable option for processors targeting laptop computers and other machines that need a low-power sleep mode. The 21264 used static memory elements because it required clock gating to keep power under control by turning off unused elements.

Unger has done a thorough, though dense, analysis of the overhead and clocking requirements of flip-flops, transparent latches, and pulsed latches [89]. Transparent latches are used in many machines. For example, IBM has been a longtime user, extensively employing a Level-Sensitive Scan Design methodology (LSSD) with transparent latches [15, 75]. Motorola uses a similar methodology on PowerPC microprocessors [2]. Silicon Graphics mixed transparent latches in speed-critical sections with flip-flops in noncritical sections on the R1000. Pulsed latches were once deemed too risky for commercial microprocessors, but have come into favor recently among some aggressive designers. AMD used the Partovi pulsed latch in the K6 microprocessor [64].

It will be interesting to watch how latches evolve in the future. As recently as 1993, some authors predicted the exclusive use of single-phase clocking [92]; that has not come true for high-performance systems. Despite the host of new latches proposed by academics [24, 96, 97], simple transparent latches and pulsed latches are likely to vie for the lead on high-speed designs, while flip-flops will be used extensively where sequencing overhead is less important.

2.5   Summary

In this chapter we have explored the commonly used memory elements for static circuits: flip-flops, transparent latches, and pulsed latches. Transparent latches can be viewed as the basic element; a flip-flop is composed of a pair of transparent latches receiving complementary clocks, while a pulsed latch is a single transparent latch receiving a pulsed clock. Table 2.1 summarizes the performance of each type of memory element. Flip-flops are the slowest. Transparent latches are good for skew-tolerant circuit design because they can hide nearly half a cycle of clock skew and permit plenty of time borrowing. Pulsed latches have the lowest sequencing overhead of any latch type, but handle less skew and time borrowing and require the most attention to min-delay problems.

Pulsed latches are a good option for high-performance designs when skews are tightly controlled and min-delay tools are good. They are conceptually simple and compatible with most design flows because they are placed in the cycle in just the same way as flip-flops. Transparent latches

Table 2.1   Static memory elements

Element	Sequencing overhead	Time borrowing	Min-delay δ_{logic}
Flip-flop	$\Delta_{CQ} + \Delta_{DC} + t_{\text{skew}}$	0	$\Delta_{CD} + t_{\text{skew}} - \delta_{CQ}$
Transparent latch	$2\Delta_{DQ}$	$\dfrac{T_c}{2} - \Delta_{DC} - t_{\text{skew}} - t_{\text{nonoverlap}}$	$\Delta_{CD} + t_{\text{skew}} - \delta_{CQ} - t_{\text{nonoverlap}}$ (in each half-cycle)
Pulsed latch	$\Delta_{DQ} + \max(0,$ $\Delta_{DC} + t_{\text{skew}} - t_{pw})$	$t_{pw} - \Delta_{DC} - t_{\text{skew}}$	$t_{pw} + \Delta_{CD} + t_{\text{skew}} - \delta_{CQ}$

are another good option for high-performance design when clock skews are more significant, time borrowing is necessary to balance logic, or the min-delay checks are inadequate. Thus, the choice between pulsed latches and transparent latches is more a matter of designer experience and judgment than of compelling theoretical advantage. Flip-flops are the slowest solution and should be reserved for systems where sequencing overhead is a minor portion of the cycle time.

The traditional pass-gate latch implementations are fast and reasonably compact. TSPC latches only require a single clock phase, but this is an illusory benefit: ultimately the designer cares about the performance, area, and power of the complete system. The extra stages and transistors in the TSPC design make it slower and more costly, so traditional designs continue to be prevalent.

2.6 Exercises

[15] **2.1** In your own words, define static and dynamic circuit families and static and dynamic memory elements. What is the difference in the usage of the words "static" and "dynamic" as applied to circuit families versus memory elements?

[20] **2.2** Consider a system with a target cycle time of 1 GHz. The clock skew budget is 150 ps. You are considering using flip-flops, transparent latches, or pulsed latches as the memory elements. How much time is available for logic within the cycle for each of the following scenarios?

(a) Flip-flops: setup time = 90 ps; clock-to-Q delay = 90 ps; hold time = 20 ps; contamination delay = 40 ps

(b) Transparent latches: setup time = 90 ps; clock-to-Q delay = 90 ps; D-to-Q delay = 70 ps; hold time = 20 ps; contamination delay = 40 ps

(c) Pulsed latches: setup time = 90 ps; clock-to-Q delay = 90 ps; D-to-Q delay = 70 ps; hold time = 20 ps; contamination delay = 40 ps. Consider using pulse widths of 180 ps and 250 ps.

[25] **2.3** Consider the paths in Figure 2.20. Using the data from Exercise 2.2, compute the following minimum and maximum delays:

(a) Flip-flops: Δ_1, δ_1.

(b) Transparent latches: Assuming 50% duty cycle clocks, compute maximum values of Δ_1, Δ_2, and $\Delta_1 + \Delta_2$; and minimum values of δ_1, δ_2, and $\delta_1 + \delta_2$. Remember to allow for time borrowing.

(c) Pulsed latches: Δ_1, δ_1. Consider using pulse widths of 180 ps and 250 ps.

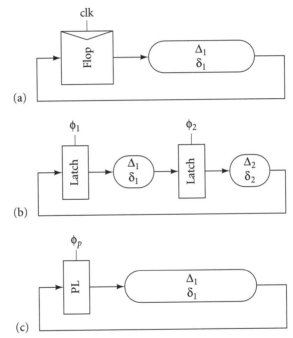

Figure 2.20 Paths using flip-flops (a), transparent latches (b), and pulsed latches (c)

[20] **2.4** Repeat part (b) of Exercise 2.3 if the duty cycle of each latch clock is d, where d is in the range of 0.3 to 0.7. Remember that the duty cycle of the clock is the fraction of time it is high. Duty cycles less than 0.5 imply nonoverlapping clocks, as shown in Figure 2.21, while duty cycles greater than 50% imply overlapping clocks.

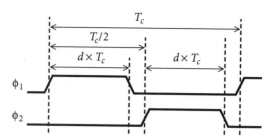

Figure 2.21 Two-phase nonoverlapping clocks with $d = 0.375$

[25] **2.5** Using the data from Exercise 2.2, compute the maximum values of $\Delta_1, \Delta_2, \Delta_3, \Delta_4, \Delta_1 + \Delta_2, \Delta_2 + \Delta_3, \Delta_1 + \Delta_2 + \Delta_3$, and $\Delta_1 + \Delta_2 + \Delta_3 + \Delta_4$ for the circuit in Figure 2.22. Assume 50% duty cycle clocks. Remember to allow for time borrowing.

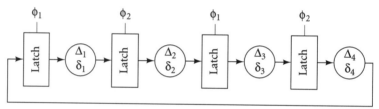

Figure 2.22 Two-cycle path using transparent latches

[10] **2.6** What are the advantages of using wide pulses for pulsed latches? What are the disadvantages of wide pulses?

[30] **2.7** Simulate the traditional static latch of Figure 2.10 in your process. Use an identical latch as the load. Make a plot of the D-to-Q delay as a function of the time from the changing data input to the falling edge of the clock. From this plot, determine the D-to-Q and setup times of the latch. Comment on how you define each of these delays.

[30] **2.8** Extend your simulation from Exercise 2.7 to determine the hold time of the latch. How do you define and measure hold time?

3
Domino Circuits

S tatic circuits built from transparent latches remove the hard edges of flip-flops, allowing the designer to hide clock skew from the critical path and use time borrowing to balance logic across pipeline stages. Traditional domino circuits are penalized by clock skew even more than are flip-flops, but by using overlapping clock phases it is possible to remove latches and thus build domino pipelines with zero sequencing overhead. This chapter describes the timing of such skew-tolerant domino pipelines and the design of individual domino gates.

3.1   Skew-Tolerant Domino Timing

This section derives the timing constraints on domino circuits that employ multiple overlapping clocks to eliminate latches and reduce sequencing overhead. The framework for understanding such systems is given the name *skew-tolerant domino* and is applicable to a variety of implementations, including many proprietary schemes developed by microprocessor designers. Once the general constraints are expressed, we explore a number of special cases that are important in practical designs. By taking advantage of the fact that clock skew tends to be less within a small region of a chip than across the entire die, we can relax some timing requirements to obtain a greater budget for global clock skew and for time borrowing. In fact, the "optimal" clocking waveforms provide more time borrowing than may be necessary, and a simplified clocking scheme with 50% duty cycle clocks may be adequate and easier to employ. Another interesting case is when the designer uses either many clock phases or few gates per cycle such that each phase controls exactly one level of logic. In this case, some of the constraints can be relaxed even further.

In general, let us consider skew-tolerant domino systems that use N overlapping clock phases. By symmetry, each phase nominally rises T_c/N after the previous phase and all phases have the same duty cycle. Each phase is high for an evaluation period t_e and low for a precharge period t_p. Waveforms for a four-phase system are illustrated in Figure 3.1.

We will assume that logic in a phase nominally begins evaluating at the latest possible rising edge of its clock and continues for T_c/N until the next phase begins. When two consecutive clock phases overlap, the logic of the first phase may run late into the time nominally dedicated to the

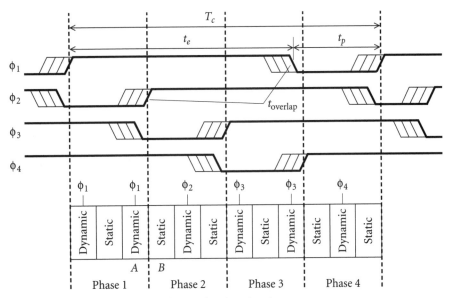

Figure 3.1   Four-phase skew-tolerant domino circuits

second phase. For example, in Figure 3.1, the second ϕ_1 domino gate consists of dynamic gate A and static gate B. Although ϕ_1 gates should nominally complete during Phase 1, this gate runs late and borrows time from Phase 2. The maximum amount of time that can be borrowed depends on the guaranteed overlap of the consecutive clock phases. This guaranteed overlap in turn depends on the nominal overlap minus the clock skew between the phases. Therefore, the nominal overlap of clock phases t_{overlap} dictates how much time borrowing and skew tolerance can be achieved in a domino system.

3.1.1   General Timing Constraints

We will analyze the general timing constraints of skew-tolerant domino by solving for this precharge period, then examining the use of the resulting overlap. Figure 3.2 illustrates the constraint on precharge time set by two consecutive domino gates in the same phase. t_p is set by the requirement that dynamic gate A must fully precharge, flip the subsequent static gate A', and bring the static gate's output below V_t by some noise margin before domino gate B reenters evaluation so that the old result from A' doesn't cause B to evaluate incorrectly. We call the time required t_{prech} and enforce a design methodology that all domino gates can precharge in this

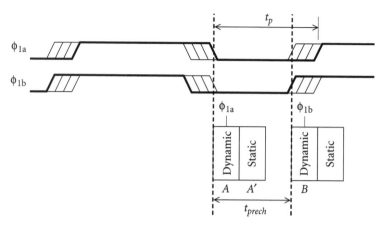

Figure 3.2   Precharge time constraint

time. The worst case occurs when ϕ_{1a} is skewed late, yet ϕ_{1b} is skewed early, reducing the effective precharge window width by t_{skew1}, which is the skew between two gates in the same phase. Therefore, we have a lower bound on t_p to guarantee proper precharge:

$$t_p = t_{prech} + t_{skew1} \tag{3.1}$$

The precharge time t_{prech} depends on the capacitive loading of each gate, so it is necessary to set an upper bound on the fanout each gate may drive. Moreover, on-chip interconnect between domino gates introduces RC delays that must not exceed the precharge time budget.

From Figure 3.1 it is clear that the nominal overlap between consecutive phases t_{overlap} is the evaluation period minus the delay between phases, $t_e - T_c/N$. Substituting $t_e = T_c - t_p$, we find

$$t_{\text{overlap}} = T_c - t_{prech} - t_{skew1} - \frac{T_c}{N} \tag{3.2}$$

This overlap has three roles. Some minimum overlap is necessary to ensure that the later phase consumes the results before the first phase precharges. This time is called t_{hold}, though it is a slightly different kind of hold time than we have seen with latches. Additional nominal overlap is necessary so that the phases still overlap by t_{hold} even when there is clock skew. The remaining overlap after accounting for hold time and clock skew is available for time borrowing. If the clock skew between consecutive phases ϕ_1 and ϕ_2 is t_{skew2}, we therefore find

$$t_{\text{overlap}} = \frac{N-1}{N} T_c - t_{\text{prech}} - t_{\text{skew1}} = t_{\text{hold}} + t_{\text{skew2}} + t_{\text{borrow}} \quad (3.3)$$

The hold time is generally a small negative number because the first dynamic gate in the later phase evaluates immediately after its rising clock edge while the precharge must ripple through both the last dynamic gate and following static gate of the first phase. Moreover, the gates are generally sized to favor the rising edge at the expense of slowing the precharge. The hold time does depend on the fanouts of each gate, so minimum and maximum fanouts must be specified in a design methodology to ensure a bound on hold time. For simplicity, we will sometimes conservatively approximate t_{hold} as zero.

Also, now that we are defining different amounts of clock skew between different pairs of clocks, we can no longer clearly indicate the amount of skew with hashed lines in the clock waveforms. Instead, we must think about which clocks are launching and receiving signals and allocate skew accordingly. We will revisit this topic in Chapter 5.

How much skew can a skew-tolerant domino pipeline tolerate? Assuming no time borrowing is used and that all parts of the chip budget the same skew, $t_{\text{skew-max}}$, we can solve Equation 3.3 to find

$$t_{\text{skew-max}} = \frac{\frac{N-1}{N} T_c - t_{\text{hold}} - t_{\text{prech}}}{2} \quad (3.4)$$

For many phases N and a long cycle T_c, this approaches $T_c/2$, which is the same limit as we found for transparent latches in Equation 2.5. Small N reduces the tolerable skew because phases are more widely separated and thus overlap less. The budget for precharge and hold time further reduces tolerable skew. The following example explores how much skew can be tolerated in a fast system.

EXAMPLE 3.1   Consider a microprocessor built from skew-tolerant domino circuits with a cycle time $T_c = 16$ and precharge time $t_{\text{prech}} = 4$ FO4 inverter delays. How much clock skew can the processor withstand if 2, 3, 4, 6, or 8 clock phases are used?

SOLUTION   Let us assume the hold time is 0. The maximum tolerable skew is computed from Equation 3.4, and the precharge period is then found using Equation 3.1. Figure 3.3 illustrates the clock waveforms, precharge period, and maximum tolerable skew for each number of

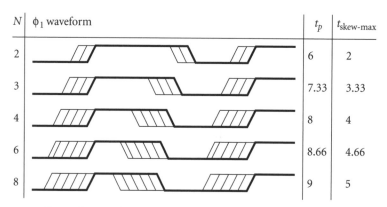

N	ϕ_1 waveform	t_p	$t_{\text{skew-max}}$
2		6	2
3		7.33	3.33
4		8	4
6		8.66	4.66
8		9	5

Figure 3.3   Skew tolerance for various numbers of clock phases

clock phases. Notice that the precharge period must lengthen with N to accommodate the larger clock skews while still providing a minimum guaranteed precharge window.   ■

In Section 5.2 we will consider various approaches for generating the clock waveforms. Ideally, t_e and the delay between phases would scale automatically as the frequency changes to provide the maximum tolerable skew at any frequency. In practice, it is often most convenient to generate fixed delays between phases, optimizing for $t_{\text{skew-max}}$ at a target operating frequency. In such a case, slowing the clock will not increase $t_{\text{skew-max}}$. Therefore, if the actual skew between any two dynamic gates exceeds $t_{\text{skew-max}}$, the circuit may fail completely, no matter how fast the gates within the pipeline evaluate and how much the clock is slowed.

3.1.2   Clock Domains

In Equation 3.4 we assumed that clock skew was equally severe everywhere. In real systems, however, we know that the skew between nearby elements, $t_{\text{skew}}^{\text{local}}$, may be much less than the skew between arbitrary elements, $t_{\text{skew}}^{\text{global}}$. We take advantage of this tighter bound on local skew to increase the tolerable global skew. We therefore partition the chip into multiple regions, called *local clock domains*, which have at most $t_{\text{skew}}^{\text{local}}$ between clocks within the domain.

If we require that all connected blocks of logic in a phase are placed within local clock domains, we obtain $t_{\text{skew1}} = t_{\text{skew}}^{\text{local}}$. We still allow arbitrary communication across the chip at phase boundaries, so we must budget $t_{skew2} = t_{skew}^{global}$. Substituting into Equation 3.3, we can solve for the maximum tolerable global skew assuming no time borrowing:

$$t_{skew\text{-}max}^{global} = \frac{N-1}{N}T_c - t_{hold} - t_{prech} - t_{skew}^{local} \tag{3.5}$$

This equation shows that reducing the local skew increases the amount of time available for global skew. In the event that local skew is tightly bounded, a second constraint must be checked on precharge that the last gate in a phase precharges before the first gate in the next phase begins evaluation a second time extremely early because of huge global skew. The analysis of this case is straightforward, but is omitted because typical chips should not experience such large global skews.

Remember that overlap can be used to provide time borrowing as well as skew tolerance; indeed, these two budgets trade off directly. Again using Equation 3.3, we can calculate the budget for time borrowing assuming fixed budgets of local and global skew:

$$t_{borrow} = \frac{N-1}{N}T_c - t_{hold} - t_{prech} - t_{skew}^{local} - t_{skew}^{global} \tag{3.6}$$

Example 3.2 illustrates the amount of time available for borrowing in an aggressive microprocessor.

EXAMPLE 3.2   Consider the microprocessor of Example 3.1 with a cycle time $T_c = 16$ and precharge time of $t_{prech} = 4$ FO4 inverter delays. Further assume the global skew budget is 2, the local skew budget is 1, and the hold time is 0 FO4 delays. How much time can be borrowed as a function of the number of clock phases?

SOLUTION   Figure 3.4 illustrates the clock waveforms, precharge period, and maximum time borrowing for various numbers of clock phases. Notice that the best clock waveforms remain the same because the clock skew is fixed, but that the amount of time borrowing available increases with N. A two-phase system offers a small amount of time borrowing, which makes balancing the pipeline somewhat easier. A four-phase design offers more than a full phase of time borrowing, granting the designer tremendous flexibility. More phases offer diminishing returns. In Chapter 5, we will find that generating the four domino clocks is relatively easy. Therefore, four-phase skew-tolerant domino is a reasonable design choice, which we will use in the circuit methodology of Chapter 4.   ∎

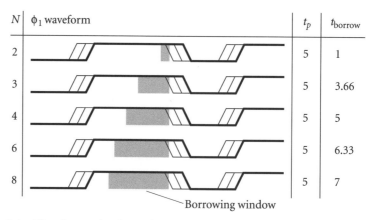

Figure 3.4 Time borrowing for various numbers of clock phases

3.1.3 Fifty-Percent Duty Cycle

Example 3.2 showed that skew-tolerant domino systems with four or more phases and reasonable bounds on clock skew may be able to borrow huge amounts of time using clock waveforms with greater than 50% duty cycle. As we will see in Chapter 5, it is possible to generate such waveforms with clock choppers, but it would be simpler to employ standard 50% duty cycle clocks, with $t_e = t_p = T_c/2$.

We have seen that the key metric for skew-tolerant domino is the overlap between phases, $t_e - T_c/N$. We know that this overlap must exceed the clock skew, hold time, and time borrowing. Substituting $t_e = T_c/2$, we find

$$t_{\text{skew-max}}^{\text{global}} + t_{\text{borrow}} = \frac{N-2}{2N} T_c - t_{\text{hold}} \tag{3.7}$$

From Equation 3.7, we see that again the overlap approaches half a cycle as the number of clocks N gets large. Of course we must use more than two clock phases to obtain overlapping 50% duty cycle clocks.

Another advantage of the 50% duty cycle waveforms is that a full half-cycle is available for precharge. This may allow more time for slow precharge or may allow the designer to tolerate more skew between precharging gates, eliminating the need to place all paths through gates of a particular phase in the same local clock domain. Of course, in a system with 50% duty cycle clocks, tighter bounds on local skew do not permit greater global skew.

3.1.4   Single Gate per Phase

In the limit of using very many clock phases or very few gates per cycle, we may consider designing with exactly one gate per phase. The precharge constraint of Equation 3.1 requires that a gate must complete precharge before the next gate in the same phase begins evaluation so that old data from the previous cycle does not interfere with operation. Skew between the two gates in the same phase is subtracted from the time available for precharge. Because there is no next gate in the same phase when we use exactly one gate per phase, we can relax the constraint. The new constraints, shown in Figure 3.5, are that the domino gate must complete precharge before the gate in the next phase reenters evaluation, and that the dynamic gate A in the current phase must precharge adequately before the current phase reenters evaluation. In a system with multiple gates in a phase, both A and B must complete precharge by the earliest skewed rising edge of ϕ_1.

As a result of these relaxed constraints, a shorter precharge time t_p may be used, permitting more global skew tolerance or time borrowing. Alternatively, for a fixed duty cycle, more time is available for precharge.

3.1.5   Min-Delay Constraints

Just as we saw in Section 2.2.4 that static circuits have hold time constraints to avoid min-delay failure, domino circuits also have constraints

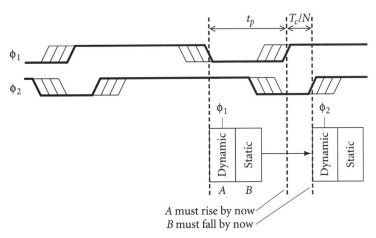

Figure 3.5   Relaxed precharge constraint assuming single gate per phase

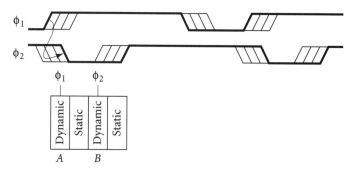

Figure 3.6   Min-delay problem in skew-tolerant domino

that data must not race through too early.[1] These constraints are identical in form to those of static logic: data departing one clocked element as early as possible must not arrive at the next clocked element until Δ_{CD} after the sampling—that is, falling—edge of the next element.

Figure 3.6 illustrates how min-delay failure could occur in skew-tolerant domino circuits by looking at the first two phases of an N-phase domino pipeline. Gate A evaluates on the rising edge of ϕ_1, which in this figure is skewed early. If the gate evaluates very quickly, data may arrive at gate B before the falling edge of ϕ_2, as indicated by the arrow. This can contaminate the previous cycle result of gate B, which should not have received the new data until after the next rising edge of ϕ_2.

The min-delay constraint determines a minimum, possibly negative, delay δ_{logic} through each phase of domino logic to prevent such race-through. The data must not arrive at the next phase until a hold time after the falling edge of the previous cycle of the next phase. This falling edge nominally occurs $T_c/N - t_p$ after the rising edge of the current phase. Moreover, skew must be budgeted because the current phase might begin early relative to the next phase. In summary:

$$\delta_{logic} \geq \Delta_{CD} + t_{skew} + \frac{T_c}{N} - t_p \tag{3.8}$$

The hold time Δ_{CD} is generally close to zero because data can safely arrive as soon as the domino gate begins precharge.

---

1. Do not confuse the min-delay hold time Δ_{CD} with t_{hold}, which is the time one phase must overlap another for the first gate of the second phase to evaluate before the last gate of the first phase precharges.

If the required δ_{logic} is negative, the circuit is guaranteed to be safe from min-delay problems. Two-phase skew-tolerant domino has severe min-delay problems because $t_p < T_c/2$, so the minimum logic contamination delay will certainly be positive. Four-phase skew-tolerant domino with 50% duty cycle clocks, however, can withstand a quarter cycle of clock skew before min-delay problems could possibly occur.

To get around the min-delay problems in two-phase skew-tolerant domino, we can introduce additional 50% duty cycle clocks at phase boundaries, as was done in Opportunistic Time-Borrowing Domino [41, 80] and shown in Figure 3.7. dlclk and dlclk_b play the roles of ϕ_1 and ϕ_2 in two-phase skew-tolerant domino, and 50% duty cycle clocks clk and clk_b are used at the beginning of each half-cycle to alleviate min-delay problems. However, this approach still requires four phases when the extra clocks are counted and does not provide as much skew tolerance or time borrowing as regular four-phase skew-tolerant domino, so it is not recommended.

3.1.6 Recommendations and Design Issues

In this section, we have explored how skew tolerance, time borrowing, and min-delay vary with the duty cycle and number of phases used in skew-tolerant domino. We have found that four-phase skew-tolerant domino with 50% duty cycle clocks is an especially attractive choice for

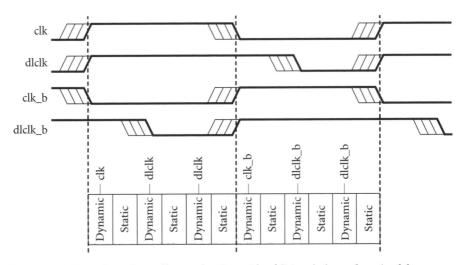

Figure 3.7 Two-phase skew-tolerant domino with additional phases for min-delay safety [41]

current designs because it provides reasonably good skew tolerance and time borrowing, is safer from min-delay problems than two-phase skew-tolerant domino, and, as we shall see in Chapter 5, uses a modest number of easy-to-generate clocks. Increasing the number of phases beyond four provides diminishing returns and increases complexity of clock generation. Therefore, we will focus on four-phase systems in the next chapter as we develop a methodology to mix static logic with four-phase skew-tolerant domino. Before moving on, however, it is worth mentioning a number of design issues faced when using skew-tolerant domino circuits.

The designer must properly balance logic among clock phases, just as with transparent latches where logic must be balanced across half-cycles. It is generally necessary to include at least one gate per phase because all of the timing derivations in this chapter have worked under such an assumption. When a large number of phases are employed, it is possible to have no logic in certain phases at the expense of skew tolerance and time borrowing; for example, an eight-phase system with no logic in every other phase is indistinguishable from a four-phase system with logic in every phase. The most common reason phases contain no logic is that the paths are short. These noncritical paths can introduce domino buffers to guarantee at least one gate per phase, or may be implemented with static logic and latches because the speed of domino is unnecessary.

In Section 3.1.2, we showed that if we could build each phase of logic in a local clock domain, we could shorten the precharge period t_p because less skew must be budgeted between gates in the same phase. Unfortunately, in some cases critical paths and floorplanning constraints may prevent a phase of logic from being entirely within a local domain. If we do not handle such cases specially, we could not take advantage of local skew to increase tolerable global skew. Two solutions are to require that the gates at the clock domain boundary precharge more quickly or to introduce an additional phase at the clock domain crossing delayed by an amount less than T_c/N.

A final issue encountered while designing skew-tolerant domino is the interface between static and domino logic. Because nonmonotonic static logic must set up before the earliest the domino clock may rise, but the domino gate may not actually begin evaluating until the latest that its clock may rise, we have introduced a hard edge at the interface and must budget clock skew. This encourages designers to build critical paths and loops entirely in domino for maximum performance. The interface will be considered in more detail in Chapter 4.

3.2 Domino Gate Design

Now that we have analyzed skew-tolerant domino timing, let us return to general issues in domino design applicable to both skew-tolerant and traditional domino circuits. In Section 1.4.1 we saw that the monotonicity rule only allowed the use of noninverting gates in a domino pipeline. To implement arbitrary logic functions, we can employ a more general logic family, dual-rail domino. We also saw that the precharge rule required us to avoid contention between pulldown and pullup transistors during precharge. We will look at the use of footed and unfooted gates to satisfy this rule. In systems that stop the clock to dynamic gates, it is necessary to use keepers to avoid floating nodes. We will look at various issues in keeper design. Once these basic issues of monotonicity, precharge, and static operation are understood, we will address robustness issues necessary for reliable operation.

3.2.1 Monotonicity and Dual-Rail Domino

The monotonicity rule says that dynamic gates should have monotonically rising inputs during evaluation; that is, the inputs should start low and stay low, start high and stay high, or start low and rise, but never start high and fall because the dynamic gate will never be able to recover from a false evaluation. Because the output of a dynamic gate is monotonically falling, it must be followed by an inverting static gate to create a monotonically rising input to the next dynamic gate. The domino gate, consisting of the dynamic/static pair, therefore always performs two inversions and thus can only implement noninverting functions. Such functions are called monotonic; they include AND and OR, but not XOR.

In traditional domino circuits, a common solution to the monotonicity problem is to structure a block of logic into a first monotonic portion followed by a second nonmonotonic portion. The first portion is implemented with domino logic, while the second portion is built with static CMOS gates. The result goes to a latch at the end of the half-cycle, and domino logic may begin again at the start of the next half-cycle. For example, a 64-bit tree adder consists of a carry generation tree followed by a final add. The carry generation to compute the carry in to each bit is monotonic, but the final add requires the nonmonotonic XOR function. Therefore, some designs perform the add in a half-cycle by building the carry logic with domino and the XOR with static CMOS at the end of the half-cycle [49]. This approach has three drawbacks. One is that mixing domino and static CMOS is not as fast as using domino everywhere. Another is that it requires traditional domino clocking techniques and thus has severe sequencing overhead. A third is that the adder is forced to fit in half of a cycle so that the nonmonotonic output arrives at the first half-cycle latch. This may limit the cycle time of the machine.

An alternative solution to monotonicity is to construct dual-rail domino circuits, also known as dynamic differential cascade voltage switch logic (DCVS) [35]. Dual-rail domino circuits accept both true and complementary versions of the inputs and produce true and complementary outputs. For example, Figure 3.8 shows a dual-rail domino AND/NAND gate, and Figure 3.9 shows an XOR/XNOR gate.

The true and complementary signals are labeled _h and _l, respectively. When a dual-rail domino gate precharges, both _h and _l outputs

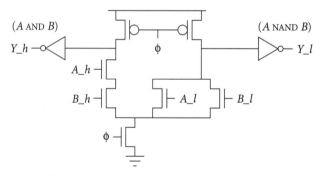

Figure 3.8 Dual-rail domino AND/NAND

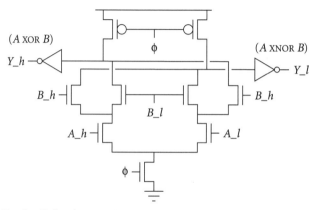

Figure 3.9 Dual-rail domino XOR/XNOR

are low, representing a quiescent state. When the gate evaluates, either _h
or _l will rise, indicating that the function is TRUE or FALSE, respectively.
The gate should never assert both _h and _l outputs high simultaneously;
this state would indicate erroneous operation of the gate.

Dual-rail domino gate design is much like single-rail design, but
requires building pulldown stacks implementing both true and complementary versions of the function. For some functions, these stacks are
completely independents; for example, in Figure 3.8 the AND function is a
regular domino AND gate using true versions of the inputs, while the
NAND function is a domino OR gate operating on complementary inputs.
For other functions, the stacks may be partially shared as illustrated in
Figure 3.9. Sharing reduces the input capacitance and thus generally
improves speed as well as area. Sharing can be determined by inspection
or by Karnaugh maps or tabular methods [12].

Dual-rail domino gates require dual-rail inputs. The gates producing
these inputs must in turn receive dual-rail inputs. This means that entire
blocks of logic must be implemented in dual-rail fashion. For example,
the 64-bit tree adder discussed earlier can be built entirely from dual-rail
domino. This improves performance by hiding the sequencing overhead
and avoiding the use of a slow static XOR, leading to a speedup of 25% or
more in aggressive systems [30]. The improvement comes at the expense
of twice as many wires to carry the dual-rail signals and greater clock
loading serving the dual-rail gates.

Dual-rail domino design is otherwise very similar to regular domino
design. In the next sections, discussions of footed and unfooted gates,
keepers, and noise margins apply equally to regular domino and dual-rail
domino logic.

3.2.2   Footed and Unfooted Gates

The precharge rule of domino gates states that there must be no active
path from the output to ground in a domino gate during precharge. Otherwise, the gate would dissipate excess power and the output would not
precharge adequately. This rule can be enforced in two ways: an explicit
transistor can be used to cut off the path to ground, or the circuit can be
timed such that paths to ground are deactivated before precharge begins.
Footed gates use the extra transistor, while unfooted gates can be faster;
these styles were shown in Figure 1.10.

Footed gates are safe to use anywhere, but unfooted gates require that all paths to ground be deactivated before the gate begins precharge. One way to guarantee this is to use inputs coming from other domino gates and to delay the precharge until the inputs have had time to precharge. For example, Figure 3.10 illustrates proper use of footed and unfooted gates. Dynamic gate A receives inputs from a static latch. The oval represents a pulldown stack of NMOS transistors implementing an arbitrary function. Because that latch output might be high during precharge, the gate needs a series evaluation transistor to avoid contention during precharge. Dynamic gate B begins precharge on the falling edge of ϕ_2 and receives inputs from a dynamic gate A that precharges on the falling edge of ϕ_1. Hence, the input will be low by the time gate B begins precharge, so no evaluation transistor is required. Gate C begins precharge at the same time as B, so its inputs are not low until partway through the precharge period. Therefore, gate C should use an evaluation transistor to avoid contention. In summary, an unfooted gate may be used when its inputs come from a domino gate that begins precharge earlier.

Note that only one transistor in each series stack must be off during precharge to avoid contention. For example, an unfooted dynamic NAND can receive some inputs from static latches so long as at least one input came from a previous phase of dynamic logic. On the other hand, an unfooted dynamic NOR gate must have all inputs low during precharge to avoid contention.

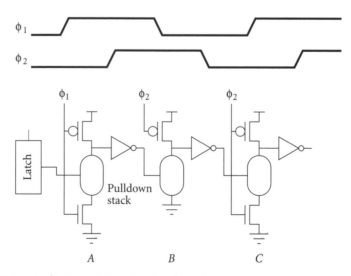

Figure 3.10   Application of footed and unfooted gates

To avoid contention in an unfooted gate completely, the gate must not begin precharge until the previous gate has had enough time to fully precharge. Guaranteeing this across process and environmental variations is difficult and makes unfooted gates hard to use. However, if the gate begins precharge while its inputs are falling but not completely low, contention will only occur for a short time. This may be acceptable in some design methodologies that sacrifice power to achieve maximum speed.

The method of logical effort [79] can be used to estimate the advantage of unfooted gates. Recall that the logical effort of a gate is the ratio of its input capacitance to that of a static CMOS inverter that can deliver the same output current. The logical effort indicates how good the gate is at driving a capacitive load; lower logical efforts correspond to faster gates. Figure 3.11 compares three dynamic inverters sized for equal output current as a static CMOS inverter. The static CMOS inverter (a) presents three units of capacitance to the A input, as indicated by the bold transistor widths. Footed dynamic inverter (b) has two units of input capacitance on the data input, meaning the logical effort is 2/3. Unfooted dynamic inverter (c) has one unit of input capacitance, meaning the logical effort is only 1/3. Footed dynamic inverter (d) uses a larger clocked evaluation transistor to allow a smaller data input transistor, achieving a logical effort of 4/9. In summary, unfooted dynamic gates have the lowest logical effort. Footed dynamic gates can approach the logical effort of unfooted gates at the expense of larger clocked transistors. Although these transistors do not load the critical path, they do increase clock loading, which increases power consumption and, indirectly, clock skew.

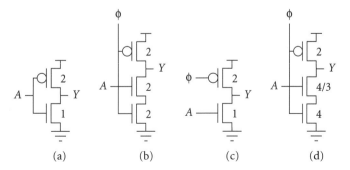

Figure 3.11   Logical effort of dynamic inverters: static CMOS (a), footed dynamic (b), unfooted dynamic (c), and footed dynamic with big foot (d)

3.2.3   Keeper Design

The dynamic gates we have shown so far have a minimum operating frequency because the output may float high while the gate is in evaluation. If left floating too long, the charge may leak away. Just as latches can be made static by placing a cross-coupled inverter pair on the storage node, dynamic gates can use a *keeper*, also called a *sustainer*, to prevent the floating node from drifting. We will see later that keepers also significantly improve the noise margin of dynamic inputs.

Figure 3.12 shows a keeper on a domino gate. The keeper is a weak PMOS transistor driven by the output inverter. When the dynamic output X stays high during evaluation, the keeper supplies a trickle of current to compensate for any leakage. When X pulls low during evaluation, the keeper turns off to avoid contention.

The first dynamic gate in a phase of skew-tolerant domino may float either high or low when the clock is stopped because the precharge transistor is turned off and the inputs fall low when the previous phase precharges. Therefore, such gates may require a full keeper built from a pair of cross-coupled inverters, as shown in Figure 3.13.

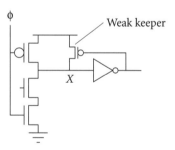

Figure 3.12   Dynamic gate with keeper

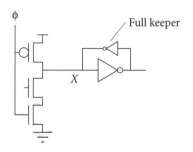

Figure 3.13   Dynamic gate with full keeper

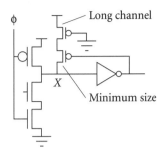

Figure 3.14   Dynamic gate with very weak keeper

Keepers should be small to minimize loading on the forward path and to avoid fighting the dynamic gate while it switches. For small dynamic gates, even minimum-sized keepers are too strong. In such a case, the channel length may be increased. Unfortunately, simply increasing the length increases the capacitive load on the critical path and thus slows the circuit. An alternative implementation of a very weak keeper is shown in Figure 3.14. The output feeds back to a minimum-size keeper. In series with the keeper is an extra transistor of minimum width and longer than minimum length that acts to reduce the current delivered by the keeper.

3.2.4   Robustness Issues

Static CMOS gates are very robust; given sufficient time, they always settle to the correct result. Domino gates are more sensitive to noise and can be irrecoverably corrupted if exposed to noise. Therefore, an essential part of domino design is a good understanding and verification of noise problems. This section surveys the noise sensitivity of dynamic inputs and outputs, then examines several sources of noise in domino circuits: charge sharing, capacitive coupling, power supply noise, alpha particles, and minority carrier injection.

Both the input and output of dynamic gates are sensitive to noise, but the mechanisms differ. In Figure 3.15, node X is the output of a dynamic gate and the input of a HI-skew static gate. It is precharged high, then may float high during evaluation. Because it is floating rather than actively driven, it is especially sensitive to noise that might cause it to fall. Moreover, the HI-skew inverter is sized to respond quickly to a falling transition on x. Therefore, its switching threshold is likely closer to 2/3 V_{DD} than 1/2 V_{DD}, leaving a smaller noise margin. Node Y is the input to the next

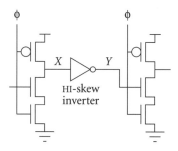

Figure 3.15   Noise-sensitive parts of domino logic

dynamic gate and output of the HI-skew inverter. It is less prone to noise because it is actively driven by the inverter. However, the noise margin of the dynamic input is only a threshold voltage V_t before the NMOS transistor begins to conduct and improperly discharge the dynamic gate.

In summary, dynamic outputs are especially noise-prone, but go to receivers with medium noise margins. Dynamic inputs are less noise-prone, but go to receivers with tiny noise margins. In the remainder of this section, we will explore the sources of noise that may impact the dynamic inputs and outputs.

Charge Sharing

Charge sharing occurs on a dynamic output when a transistor turns on and transfers charge between two floating capacitors. For example, consider the dynamic NAND gate in Figure 3.16. The capacitor C_y on the internal node represents parasitic diffusion capacitance, while the capacitor C_x on the output represents both diffusion capacitance and the output load. Initially, suppose the gate is in evaluation, the output X is precharged to V_{DD}, but that internal node Y was discharged to GND. When the top NMOS transistor turns on, the output is not logically supposed to change because the other NAND input is low. However, charge redistributes between C_x and C_y until the voltages equalize.[2] This charge sharing causes the output to droop while the internal node rises. If the output droops too far, an incorrect value is produced. The final voltage on the output is set by the capacitive voltage divider equation:

$$V_x = \frac{C_x}{C_x + C_y} V_{DD} \tag{3.9}$$

2. Or at least until C_y rises to within a threshold drop of V_{DD}.

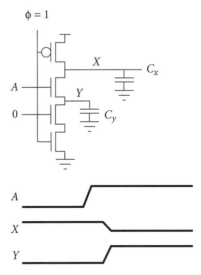

Figure 3.16 Charge-sharing example

Charge sharing is minimized by making the internal diffusion capacitances small compared to the load capacitance. The diffusion capacitance depends on layout, so it is important to carefully lay out dynamic gates. Driving larger load capacitances also reduces charge sharing, but of course increasing the load excessively slows the dynamic gate.

Even after careful layout, charge sharing is often unacceptably great in complex domino gates with many internal nodes. In such a case, charge sharing can be alleviated by precharging internal nodes with a secondary precharge transistor, as shown for a complex AND-OR-INVERT (AOI) gate in Figure 3.17. Each precharge device adds extra diffusion capacitance that slightly slows the gate, so it is best to only precharge the smallest number of internal nodes necessary to satisfy noise margins. A guideline is that

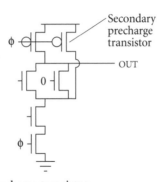

Figure 3.17 Secondary precharge transistor

precharging every other node is generally sufficient to keep charge-sharing noise to about 10% of the supply voltage for most gates. Be careful of contention if precharging internal nodes in unfooted dynamic gates.

A related form of charge sharing occurs when dynamic gates drive pass transistors, as shown in Figure 3.18. When the pass transistor turns on, charge is shared between the dynamic node X and the pass transistor output Y. Therefore, dynamic gates should not drive the source/drain inputs of pass transistors.

Capacitive Coupling

Capacitive coupling, also known as *crosstalk,* is a major component of noise on both inputs and outputs of dynamic gates. Wires adjacent to a domino gate may have capacitance to the dynamic gate input or output. The adjacent wire is called the *aggressor* or *perpetrator,* while the dynamic input or output is the *victim.* When the aggressor switches, it tends to drag the victim along, injecting noise onto the victim. Floating dynamic outputs are especially susceptible to coupling. Dynamic inputs must also be checked because of their tiny noise margin, but receive less coupling because they are actively driven and the driver fights against the coupling.

The coupling onto a dynamic output is modeled in Figure 3.19. The victim has C_{couple} coupling capacitance to the aggressor and C_{victim} capacitance to other nodes such as the substrate that are not switching. When the aggressor line falls, the victim line will tend to fall too. As with charge

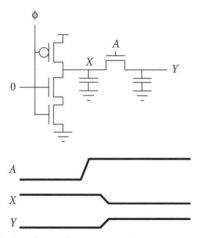

Figure 3.18 Charge sharing through a pass transistor

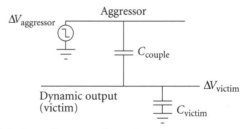

Figure 3.19   Model of coupling onto dynamic output

sharing, we have a simple capacitive voltage divider, so the noise ΔV_{victim} on the victim is

$$\Delta V_{\text{victim}} = -\frac{C_{\text{couple}}}{C_{\text{couple}} + C_{\text{victim}}} \Delta V_{\text{aggressor}} \tag{3.10}$$

The coupling onto a dynamic input is modeled in Figure 3.20. The victim is sensitive to a rising aggressor, which could drag the victim high enough to turn on the dynamic gate. This time, the resistance of the aggressor and victim drivers is important. This resistance can be computed from the linear I-V characteristics of the driver transistor. If the victim has a very strong—that is, low-resistance—driver, it will be virtually immune to coupling because its driver will keep the victim low. The time constant ratio k of the aggressor to victim determines how fast the victim can respond relative to the rate at which the aggressor injects noise [91]:

$$V_{\text{victim}} = \frac{C_{\text{couple}}}{C_{\text{couple}} + C_{\text{victim}}} \frac{1}{1 + k} V_{DD} \tag{3.11}$$

where the time constant ratio is

$$k = \frac{\tau_{\text{aggressor}}}{\tau_{\text{victim}}} = \frac{R_{\text{aggressor}}(C_{\text{aggressor}} + C_{\text{couple}})}{R_{\text{victim}}(C_{\text{victim}} + C_{\text{couple}})} \tag{3.12}$$

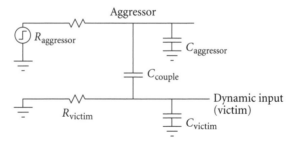

Figure 3.20   Model of coupling onto dynamic output

Coupling is an increasing problem as the height-to-separation ratio of wires increases in finer geometry processes, increasing the ratio of C_{couple} to other capacitances. For example, even in a 0.8-micron process, over 50% of the capacitance of a minimum pitch metal 2 wire is to adjacent lines. It is important not to be overly pessimistic about coupling lest design of domino logic become impossible. For example, we can take advantage of the fact that a victim is only sensitive to noise while in evaluation to avoid factoring in coupling noise from aggressors that switch in a different phase.

Coupling problems can be reduced by increasing the spacing between the aggressor and victim or by shielding the victim with narrow ground lines. In dual-rail domino, proper routing can also reduce coupling problems, as illustrated in Figure 3.21. If signals are routed as shown on the left, a victim may be susceptible to aggressors on both sides. If the signals are interdigitated as shown on the right, the victim will never see more than one aggressor switching at a time because either the _h or _l rail, but not both, will switch. Other approaches to coupling noise reduction include recoding busses as one-hot for fewer transitions [34] and inserting extra static inverters at the beginning and end of long lines to filter noise [62].

Power Supply Noise

Another source of noise in dynamic gates comes from power supply variation across the die. Suppose a static inverter drives a dynamic input and that the ground potential of the inverter is higher than that at the dynamic gate due to IR drops across the ground network, as shown in Figure 3.22. If the ground difference exceeds V_t, the dynamic gate will turn on and incorrectly evaluate. Of course, power supply noise impacts dynamic outputs as well, though the noise margins are greater.

Power supply noise is especially insidious because of the trends toward lower voltage and higher power. Higher power increases the supply cur-

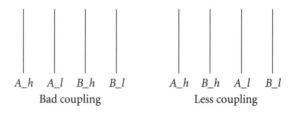

Figure 3.21   Reducing coupling in dual-rail domino

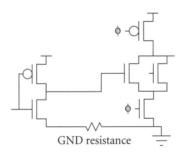

GND resistance

Figure 3.22   Power supply noise at a dynamic input

rent. Lower voltage also increases the supply current if power were held constant. Hence, the current is increasing quadratically. For a fixed current, the IR drop through a power supply wire increases as a fraction of the supply voltage when voltages drop. Therefore, power supply noise as a fraction of the supply voltage is getting worse cubically if supply resistance does not change! To combat this, designers are dedicating increasing amounts of metal to the power supply network to reduce the resistance. Some chips place solder bumps carrying power and ground approximately every millimeter over the surface of the die to minimize the length and resistance of supply lines. The DEC Alpha 21264 lacked such solder bump technology so instead dedicated entire planes of metal to V_{DD} and GND [22] in much the same way as multilayer printed circuit boards use power and ground planes. The amount of supply noise is a trade-off between cost and performance. Typically enough metal resources are dedicated to the supply to keep IR drops down to 10% across the chip [52].

Another source of power supply noise is di/dt noise from changing supply current. Power supply pins have inductance and thus higher impedance at high frequencies; they cannot supply all of the instantaneous current required by gates. Instead, this current is drawn from on-chip bypass capacitance. Idle gates present substantial bypass capacitance, but high-speed designs often must add extra explicit bypass capacitors in empty parts of the chip, especially near large clock drivers. The greatest supply noise occurs when the clock switches. Therefore, multiphase domino clocking reduces peak noise by spreading the current spikes across the cycle.

Alpha Particles

Alpha particles are helium nuclei (two protons and two neutrons) emitted from radioactive decay. They occasionally crash into chips and leave a trail of electron-hole pairs in the silicon substrate [40]. These carriers may be collected onto floating nodes, disturbing the voltage of dynamic outputs. Such disturbances are called *soft errors*. To minimize this change, the energy held on the capacitance of the floating node must be much greater than the energy in the alpha particle. Typical design rules call for a minimum of several femtofarads of capacitance on any dynamic node [66]. Therefore, alpha particle hits are only a problem for very small dynamic gates and SRAMs.

Interestingly, lead solder is a common source of alpha particles. Chips that use C4 solder bump packing technology may have hundreds of lead solder bumps covering the die. Placing memory arrays and sensitive dynamic gates under these bumps may be hazardous! It is rumored that some companies have cornered the market on Roman lead, which has naturally decayed and thus has lower alpha particle emissions!

Minority Carrier Injection

Certain circuits such as I/O drivers powering inductive loads are prone to ringing, which may send the output below GND. In such a case, the drain-substrate junction becomes forward biased and injects charge into the substrate. This charge may be collected on nearby dynamic nodes, corrupting the value. To prevent this, circuits prone to charge injection should be placed far away from dynamic logic. They should also be surrounded by guard rings [26]. Charge injection is only a concern with special-purpose circuits and therefore is not part of the noise budget of most dynamic gates.

Noise Feedthrough

When the input of a gate is near its noise margin, the output voltage will not be at the rail. Therefore, the input of the next gate will see some noise; this is called *noise feedthrough* or *residual noise*. Indeed, the noise margin of a gate is defined by this feedthrough: it is the point at which the noise slope of the transfer function is −1 so that the marginal increase in noise feedthrough to the next gate equals the marginal increase in noise margin of the current gate.

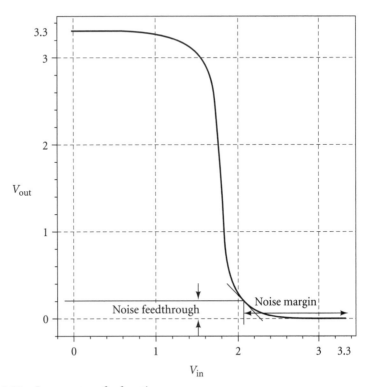

Figure 3.23   Inverter transfer function

Figure 3.23 shows the transfer function of a HI-skew inverter using a PMOS transistor four times as large as the NMOS transistor. Because we are using the inverter after a dynamic gate, we are concerned about the high input noise margin, the amount the dynamic output can droop before the HI-skew inverter no longer produces a valid 0. This is determined by the unity-gain point marked on the transfer curve. Notice that at this point, the output is not quite zero. The nonzero voltage shows up as noise at the sensitive input of the next dynamic gate. In this case, the noise margin is 1.22 volts (37% of the supply) and the noise feedthrough is 0.23 volts (7% of the supply).

Summary of the Noise Budget

We have seen that domino logic is subject to numerous sources of noise. Dynamic outputs are especially noise-prone, but drive static gates with reasonably good noise margins. Dynamic inputs experience less noise, but have a noise margin of only about V_t. Most noise sources scale with the power supply voltage, so noise budgets expressed as a fraction of the supply voltage tend not to change very much with feature size and supply voltage. In this section, we will create some sample noise budgets and see that each source of noise must be tightly bounded.

For the sake of concreteness, let us develop our budget using the Hewlett-Packard CMOS14 0.6-micron 3.3v process. As we saw in Figure 3.23, the dynamic output drives a HI-skew static inverter with a noise margin of 37% of the supply voltage, causing a noise feedthrough of 7%. We assume the dynamic gate has a small keeper and thus achieves a noise margin of 19% with a feedthrough of 5%. These noise margins are measured in the worst-case process corners: SF for the HI-skew inverter and FS for the dynamic input.

Table 3.1 shows a sample noise budget for dynamic inputs and outputs. It assumes that dynamic inputs and outputs are kept within local blocks and therefore see less power supply noise than is found across the entire die.

The sample budget shows that very little coupling noise can be tolerated. This can make domino design costly because signals must be shielded or widely spaced to avoid coupling. However, most circuits do not experience the worst case of all noise sources simultaneously. For example, dynamic NOR gates are not susceptible to charge sharing. Therefore, a less pessimistic noise verification tool can check each circuit individually to allow greater coupling noise on nodes that experience less of other noise sources. Moreover, dynamic logic is sensitive to the duration as well as the magnitude of noise; for example, larger peak coupling noise could be permitted if the peak coupling occurred for only a fraction of a gate delay [51].

Similarly, the minimum noise margins for dynamic outputs and inputs are determined by different process corners. For example, when

Table 3.1 Sample noise budget (percentage of power supply voltage)

Source	Dynamic output	Dynamic input
Charge sharing	10	n/a
Coupling	17	7
Supply noise	5	5
Feedthrough	5	7
Total	37	19

the dynamic input is at its minimum noise margin in its least favorable process corner and produces the most feedthrough noise, the dynamic output will be in its most favorable corner and will be able to tolerate more noise, somewhat alleviating the feedthrough noise problem.

A common mistake is to ignore the other sources of noise when focusing on a particular problem. For example, a designer may simulate charge sharing and conclude that no secondary precharge transistor is necessary because charge sharing that causes a 30% output voltage droop on a dynamic gate does not trip the subsequent static inverter in simulation. Of course, when other sources of noise such as coupling and supply noise are introduced, the circuit will fail.

Domino logic allows a trade-off between noise margins and performance. If the dynamic output has inadequate noise margin, it can be improved by using a lower-skew static gate; however, the gate will not respond as quickly to falling transitions, so the circuit will be slower. If the dynamic input has inadequate noise margin, it can be improved by using a bigger keeper. The keeper is especially effective because it is fully on while the NMOS pulldown transistor is just barely on when the dynamic input voltage is slightly above V_t. For example, the keeper in the sample noise budget was 1/5 the size of the input transistor. Doubling the keeper size increases the dynamic input noise margin from 19% to 27% of the supply voltage. The keeper also slightly improves the noise margin of the dynamic output. However, increasing the keeper size slows the falling transition of the dynamic output because of contention and increased loading.

3.3   Historical Perspective

The motivation for using dynamic logic has changed strikingly over the past two decades. In the NMOS days of the 1970s, dynamic logic was used to reduce power consumption inherent in NMOS logic [60, 65] and to save area on precharged busses [56]. Domino logic was proposed for CMOS as "high-speed compact" logic [49]. The risks associated with the tight domino noise margins deterred use in commercial products. Moreover, critical paths could often be solved by using more parallel, area-intensive circuits; for example, a designer could choose a static CMOS carry look-ahead adder instead of using domino logic for a ripple carry adder.

The relative importance of area, power, and speed shifted by the early 1990s. Area, once the primary limitation of CMOS circuit design, becomes cheaper with each generation of process. The trends toward shorter cycles that we observed in Section 1.2 have pushed logic to use the fastest, most parallel algorithms available because area is so inexpensive. To achieve even shorter cycle times as measured in gate delays, processors had to employ dynamic logic. For example, designers of the HP-PA8000 micro-processor found that extensive use of dynamic logic (40% of the gates were dynamic) was a key factor in achieving desired performance [20, 21]. Power consumption and area ceased to be motivations for using dynamic gates. Indeed, the power consumption of domino logic is much higher than static logic because power in logic circuits is dominated by switching the clock capacitance [25]. Trends toward more dual-rail dom-ino actually increase circuit area as well!

A key requirement for using domino logic is the thorough under-standing and checking of noise budgets. DEC has a particularly well-established and documented domino circuit methodology [26]. Remark-ably, DEC did not use keepers on dynamic gates in the 21164, despite the fact that threshold voltages were an aggressively low 0.35v! However, keepers ultimately became necessary in the 21264 to offset NMOS leakage current [3]. The use of domino has mostly been restricted to custom, highly optimized chips such as microprocessors because guaranteeing sig-nal integrity is beyond the capability of current turnkey synthesis and place-and-route tools used by most ASIC designers. IBM's static noise analysis tool [76] is an interesting approach to the problem of automatic noise analysis.

As cycle times measured in gate delays have decreased, the sequencing overhead of domino logic has become a greater problem. Williams observed that latches were unnecessary between blocks of domino logic so long as one block could evaluate before the next began precharge. He used this observation to construct "zero-overhead" asynchronous circuits [93, 94]. Synchronous designers began to adopt the ideas of skew-tolerant domino by the mid-1990s. Hewlett-Packard used a "pipeline latch" in the PA7100 floating-point multiply-accumulate (FMAC) path to soften the clock edges and reduce the impact of clock skew [33]. Intel used "Oppor-tunistic Time-Borrowing (OTB) Domino" on the Itanium Processor. OTB is a variant of two-phase skew-tolerant domino coinvented by this author [41, 80]. Another Intel team has made an extensive effort to develop self-resetting domino [11], an interesting and important technique that is beyond the scope of this book. DEC used overlapping clocks to eliminate latches between domino blocks in the ALU and select other paths of the Alpha 21164 [6]. Sun also used a similar technique, which they called "delayed reset domino," in the UltraSparc [53]. Most microprocessor companies, including Advanced Micro Devices, Silicon Graphics, Motorola, Hewlett-Packard, Sun Microsystems, and IBM, have developed the ideas of skew-tolerant domino, though have generally kept their use as a trade secret.

3.4 Summary

Domino gates have become very popular because they are the only proven and widely applicable circuit family that offers significant speedup over static CMOS in commercial designs, providing a 1.5 to 2 times advantage in raw gate delay. However, speed is determined not just by the raw delay of gates, but by the overall performance of the system. For example, traditional domino sacrificed much of the speed of gates to higher sequencing overhead. As cycle times continue to shrink, the sequencing overhead of traditional domino circuits increases and skew-tolerant domino techniques become more important. Skew-tolerant domino uses overlapping clocks to eliminate latches and remove the three sources of sequencing overhead that plague traditional domino: clock skew, latch delay, and unbalanced logic. The overlap between clock phases determines the sum of the skew tolerance and time borrowing. Systems with better bounds on clock skew can therefore perform more time borrowing to balance logic between pipeline stages. Increasing the number of clock phases increases the overlap, but also increases the complexity of local clock generation and distribution. Four-phase skew-tolerant domino, using four 50% duty cycle clocks in quadrature, is a particularly interesting design point because it provides a quarter cycle of overlap while minimizing the complexity of clock generation.

Domino gates are very susceptible to noise. Dynamic inputs have a very low noise margin and are especially impacted by coupling and power supply noise. Dynamic outputs have a larger noise margin, but are impacted by charge sharing as well. Domino gates should be checked for the following electrical rules:

- Neither inputs nor outputs of dynamic gates should connect to diffusion terminals of pass transistors.

- Nodes that may float indefinitely must be held by keepers.

- Coupling onto dynamic inputs and outputs must be limited.

- Noise margins on dynamic inputs and outputs must be checked.

The next chapter will further explore the use of skew-tolerant domino in the context of an entire system, describing a methodology compatible with other circuit techniques but that still maintains low sequencing overhead.

3.5   Exercises

[15]   **3.1**   Sketch a diagram like Figure 3.1 illustrating a six-phase domino pipeline with 50% duty cycle clocks and one domino gate per clock phase. Indicate clock skew of one-sixth of the cycle.

[15]   **3.2**   A domino gate has an evaluation time of 100 ps and a precharge time of 200 ps. If there is 50 ps of skew between the clock controlling the gate and its successors in the same phase, what is the minimum time t_p that the clock must be low?

[20]   **3.3**   Repeat Example 3.1 if the cycle time is 12 FO4 delays and the precharge time is 3 FO4 delays.

[20]   **3.4**   Repeat Example 3.2 if the cycle time is 12 FO4 delays and the precharge time is 3 FO4 delays.

[15]   **3.5**   A four-phase skew-tolerant domino pipeline runs at 800 MHz in a 0.18-micron process with a 60 ps FO4 delay. You can adjust the duty cycle of the clocks for best performance. If you allow a precharge time of 5 FO4 delays and a hold time of 1 FO4 delay, when there is 50 ps of local clock

skew, how much global skew can you tolerate? If the actual global skew is 200 ps, how much time borrowing can you permit?

[30] **3.6** Repeat Exercise 3.5 if you design to guarantee exactly one domino gate per clock phase.

[20] **3.7** A four-phase skew-tolerant domino pipeline runs at 1.25 GHz using 50% duty cycle clocks. The required overlap between phases is $t_{hold} = -15$ ps. Each domino gate has a contamination delay of 35 ps and a hold time Δ_{CD} of −10 ps. How much clock skew can the pipeline withstand before one gate might precharge before its successor could consume the result? How much clock skew can the pipeline withstand before min-delay problems might occur? In summary, how much clock skew can the system withstand?

[20] **3.8** Sketch transistor-level implementations of the following footed dual-rail dynamic gates:

    (a) OR/NOR

    (b) AND-OR-INVERT (AOI)

    (c) three-input MAJORITY (output TRUE if at least two inputs are TRUE)

    (d) three-input XOR

[20] **3.9** Sketch transistor-level implementations of the following footed dynamic gates. Label each NMOS transistor with the appropriate width to provide the same output drive as a unit inverter (see Figure 3.11). Select the PMOS transistor width for half the output drive as the pulldown stack. Estimate the logical effort of each data input to the gate.

    (a) NAND2

    (b) NAND3

    (c) NOR2

    (d) NOR3

    (e) AND-OR-INVERT (AOI)

[15]   **3.10** Repeat Exercise 3.9 for unfooted dynamic gates.

[30]   **3.11** Make plots of evaluation time and precharge time for the domino
buffer in Figure 3.24 as a function of the precharge transistor size P. The
transistor and load sizes have been selected to provide a stage effort of
about 4. Use step inputs. Measure evaluation time to 50% output of the
static inverter when ϕ is already high and A rises. Measure precharge time
from the falling edge of ϕ to the static inverter output Y dropping to 10%
of V_{DD}. Use your favorite process, environment, and SPICE simulator. Let
the dimensions be in units of 10 microns of gate width. What value of P
would you select for general application?

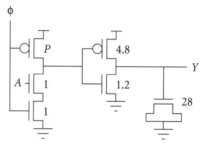

Figure 3.24   Domino buffer for simulation of precharge transistor size

[30]   **3.12** Make plots of evaluation time and input noise margin for the domino buffer in Figure 3.25 as a function of the keeper transistor size k. Use
step inputs. Measure evaluation time to 50% output of the static inverter
when ϕ is already high and A rises. Measure noise margin at the unity
gain point of the output Y. Use your favorite process, environment, and
SPICE simulator. Let the dimensions be in units of 10 microns of gate
width. What value of k would you select for general application?

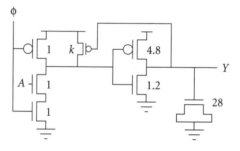

Figure 3.25   Domino buffer for simulation of keeper size

[30]   **3.13** Design a dynamic footed AOAOAOI gate to compute $\overline{B(C + D(E + F(G + H)))}$. Choose the transistor sizes to have a maximum of 20 microns of gate width on any input. The gate should drive an inverter with a total of $20h$ microns of gate width. Simulate it in SPICE, being sure to include AS, AD, PS, and PD parameters to specify diffusion parasitics. Find the worst-case charge-sharing noise on the output for $h = 0, 1, 2, 4,$ and 8. How does the noise depend on h? Why? Add secondary precharge transistors to precharge every other internal node. Repeat your charge-sharing measurements. Explain your observations.

[30]   **3.14** Simulate capacitive coupling between two metal lines. Each line has a capacitance to ground of 0.1 fF/micron and a capacitance to the adjacent line of 0.2 fF/micron. The aggressor's driver is a falling voltage step with an effective resistance of 100 Ω. The victim is a dynamic node; the keeper has an effective resistance of R. Plot the peak coupling noise versus R for 100-micron and 1 mm line lengths. How do your results compare with the predictions of Equation 3.11?

[25]   **3.15** Simulate the DC transfer characteristics of an inverter in your process, using a P/N ratio of 2. Find the unity gain points on the transfer function. Measure V_{in-l} and V_{in-h}, the input voltages at the low and high unity gain points; and V_{out-l} and V_{out-h}, the output voltages at these points. What are the high and low noise margins for your inverter?

[25]   **3.16** Repeat the simulation of Exercise 3.15 with a P/N ratio of γ. What value of γ gives equal high and low noise margins in your process?

[25]   **3.17** Identify potential noise problems in the circuit in Figure 3.26. Draw an improved circuit with reduced noise risk.

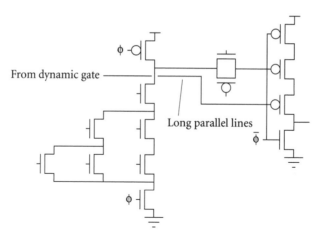

Figure 3.26   Noise-prone circuit

[15]   **3.18** An early stepping of a well-known microprocessor suffered unreliable operation due to noise. The problem was traced to a path between two widely separated units. The receiving unit used a transmission-gate latch, as shown in Figure 3.27(a). The problem could be fixed by substituting a different transparent latch, shown in Figure 3.27(b). Explain why the noise problem might occur and how the input noise margins of each latch compare.

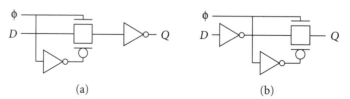

(a)   (b)

Figure 3.27   Path between units with noise problem: inverter after transmission gate (a) and inverter before transmission gate (b)

4
Circuit Methodology

In this chapter, we will develop a skew-tolerant circuit design methodology. There are many aspects to skew-tolerant circuit design, including the interface of static and domino logic, clocking, and testability.
Unless you consider all aspects simultaneously, it is easy to optimize one
aspect at the expense of the others, resulting in a system that is slower
overall or more difficult to build. Our objective is a coherent approach to
combine domino gates, transparent latches, and pulsed latches, while
providing simple clocking, easy testability, and robust operation. The
guidelines presented are self-consistent and support the design and verification of fast systems, but are not the only reasonable choices.

Our design methodology contains definitions and guidelines. Definitions provide a common set of terms that enable tools to parse the design.
Guidelines should be followed for easy, robust design. They should only
be violated after being fully understood. During a design review, guideline violations should be discussed.

We will emphasize circuit design in this chapter while deferring discussion of the clock network until the next chapter. Of course, circuit
design and clocking are intimately related, so this methodology must
make assumptions about the clocking. In particular, we assume that we
are provided four overlapping clock phases with 50% duty cycles. These
clocks will be used for both skew-tolerant domino and static latches.

DEFINITION 1   The clocks are named ϕ_1, ϕ_2, ϕ_3, and ϕ_4. Their nominal waveforms are
shown in Figure 4.1.

The clocks are locally generated from a single global clock gclk. ϕ_1 and
ϕ_3 are logically true and complementary versions of gclk. ϕ_2 and ϕ_4 are

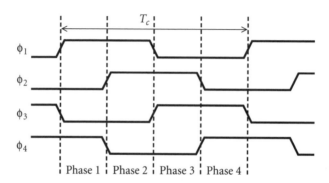

Figure 4.1   Four-phase clock waveforms

versions of ϕ_1 and ϕ_3 nominally delayed by a quarter cycle. The clocks may be operated at reduced frequency or may even be stopped while gclk is low for testability or to save power.

The methodology primarily supports four-phase skew-tolerant domino, pulsed latches, and ϕ_1 and ϕ_3 transparent latches. Other latch phases are occasionally used when interfacing static and domino logic. It is recommended, but not required, that you choose either transparent latches or pulsed latches as the primary static latch to simplify design.

4.1 Static/Domino Interface

In the previous chapters, we have analyzed systems built from static CMOS latches and logic and systems built from skew-tolerant domino. In this section, we discuss how to interface static logic into domino paths and domino results back into static logic. We focus on static logic using transparent latches and pulsed latches because flip-flops are not tolerant of skew. We develop a set of "timing types" that determine when signals are valid and allow checking that circuits are correctly connected. The guidelines emphasize performance at the cost of checking min-delay conditions.

4.1.1 Latch Placement

Before discussing the static/domino interface, we must decide where static latches belong in the cycle. Of course, pulsed latches are placed at the beginning of a cycle when the pulse occurs and domino gates are assigned to nominally evaluate during one of the four phases. Transparent latches pose a dilemma, however: should they be placed at the beginning, middle, or end of their half-cycle? We will look at each option, then conclude that placement scarcely matters in latch-based loops because the latches adjust themselves as early as possible in the cycle depending on data arrival times. Therefore, we might as well nominally place the latch at the beginning of the cycle.

In Section 1.3, we argued that latches could tolerate clock skew if they were placed in the middle of their half-cycle such that skew disturbing the latch transparency at the beginning or end of the half-cycle has no effect. Placing latches in the middle of the half-cycle presents two difficulties.

One is that the designer must determine where the "middle" is in a half-cycle of logic. Another is that designers often construct logic with a minimum number of signals crossing cycle boundaries because the cycle boundaries also represent partitions of a large system between modules and designers. Within the cycle, a larger number of intermediate results may be generated. If latches were placed in the middle of a half-cycle, more latches might be required.

The alternatives are to place latches at the beginning or end of each half-cycle. Designers who are accustomed to thinking about latches as memory elements often like to place the latch at the end of the half-cycle [66] to "remember" the result of the computation in the half-cycle. Recall, however, from Section 2.1.1 that the purpose of a latch is not to remember information but rather to retard fast signals from racing ahead and corrupting information in the next cycle. Fast signals arrive early, so the function of a latch is to remain opaque until the signals are allowed to enter the half-cycle. From this point of view, it makes sense for latches to be placed at the beginning of the half-cycle. Data that is not critical arrives at the latch before it becomes transparent. Time borrowing occurs when data arrives late and propagates through the latch while it is transparent. This is exactly the behavior observed when using a functional simulator that assumes zero delay through gates. Thinking of latches as belonging at the beginning of half-cycles is also convenient because it emphasizes the similarities of transparent latches and pulsed latches, works well with scan (Section 4.3), and matches the results of timing analysis (Chapter 6).

At first glance, it appears wasteful for data to nominally arrive at a latch while the latch is still opaque. The data will be delayed waiting for the latch to become transparent. Does this introduce sequencing overhead? Interestingly, the answer is no. Noncritical signals do arrive before the latch is transparent and are slowed to stay synchronized. Critical signals, however, are late and thus arrive while the latch is transparent. For example, Figure 4.2 shows the arrival times of noncritical and critical signals. We assume paths starting from the ϕ_1 latch depart at the latest the latch might become transparent in the event of clock skew. The top path is less than a half-cycle, so it must wait for the ϕ_3 latch to become transparent. This introduces sequencing overhead but is unimportant because the path was faster than necessary. The middle path is exactly a half-cycle, so it arrives at the ϕ_3 latch just as the latch has become transparent. There-

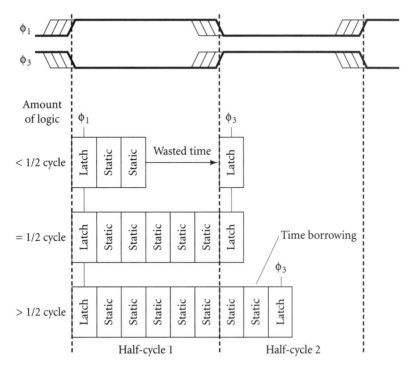

Figure 4.2   Latch placement and sequencing overhead

fore, no time is wasted. The bottom path is more than a half-cycle, so it borrows time into the second half-cycle. This time borrowing effectively pushes the ϕ_3 latch later into the second half-cycle. The maximum amount of time borrowing is set by the fact that data must arrive at the latch by a setup time before the earliest skewed falling edge of the clock. In summary, we can think of latches being nominally placed at the beginning of each half-cycle. Critical paths incur no overhead from clock skew or setup time. Time borrowing can push the actual location of the latch further into the half-cycle.

Even if latches were placed somewhere other than the beginning of a half-cycle, during operation they adjust themselves so that data arrives before the latch becomes transparent unless the logic is slow enough that time borrowing is necessary. For example, Figure 4.3 illustrates a path consisting of a loop through two latches L_2 and L_3. The loop also receives an external input from a flip-flop F_1. The propagation delays through the combinational logic (CL) are $\Delta 1$, $\Delta 2$, and $\Delta 3$. The departure times from each element, that is, the time at which data begins propagating through the latch, are D_1, D_2, and D_3. The latches are nominally placed at the end

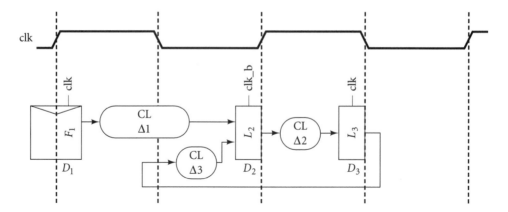

Figure 4.3　Path illustrating time borrowing and hard edges

of the half-cycles in the pipeline diagram, but we will see that if the combinational logic delays are short compared to the cycle time, data will arrive at latches before they become transparent, so the latches will effectively operate at the beginning of the half-cycles.

Figure 4.4 illustrates the circuit timing assuming $\Delta 1 = 0.2$ ns, $\Delta 2 = 0.6$ ns, and $\Delta 3 = 0.15$ ns. The three diagrams correspond to cycle times T_c of 1.5 ns, 1.0 ns, and 0.75 ns. The vertical tic marks delineate half-cycle boundaries. The loop formed by the two latches is rolled into a circle; the circumference of the circle represents the cycle time. The external input from the flip-flop is represented by the horizontal line. The heavy lines correspond to times when blocks of combinational logic are operating and the departure times from each latch are indicated on the diagram. In each case, data departs the flip-flop on the rising edge of the clock. When the cycle time is 1.5 ns, data arrives at the input of each latch before the latch becomes transparent. Therefore, the data departs the latch at the beginning of the half-cycle when the latch becomes transparent, even though the pipeline diagram had planned latches at the end of the half-cycle. When the cycle time drops to 1.0 ns, $\Delta 2$ exceeds half of the cycle time, so data departs L_3 after the clock edge. This indicates that time borrowing is taking place. When the cycle time drops to 0.75 ns, even more time borrowing is required. The arcs formed by $\Delta 2$ and $\Delta 3$ sum to the entire circumference of the circle, indicating that the system is operating at the minimum possible cycle time.

Figure 4.5 is a similar diagram showing how operation changes when $\Delta 1$ increases to 0.8 ns. Even when the cycle time is 1.5 ns, $\Delta 1$ is so long

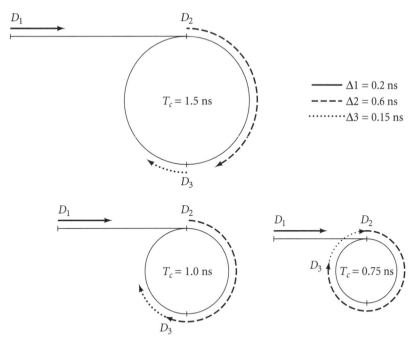

Figure 4.4 Timing diagrams with fast Δ1 for various cycle times

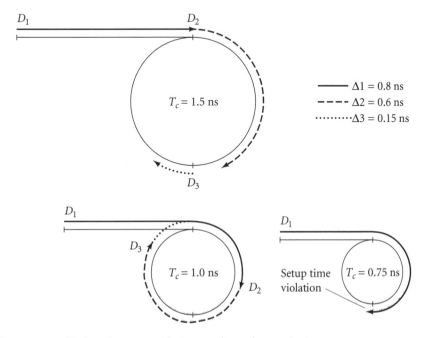

Figure 4.5 Timing diagrams with slow Δ1 for various cycle times

that data does not depart L_2 until after the clock edge. The hard edge imposed by the flip-flop determines the departure time from the latch. $\Delta 2$ is short enough, however, that no time borrowing is required through L_3. When the cycle time drops to 1.0 ns, time borrowing is required through both latches. At a cycle time of 0.75 ns, $\Delta 1$ is longer than a cycle and the setup time at L_2 is violated. Therefore, the circuit will not operate, even though the delay around the loop $\Delta 2 + \Delta 3$ exactly fits in the cycle time.

In summary, loops of logic and transparent latches automatically adjust themselves so that data departs the latches at the beginning of each half-cycle unless time borrowing is necessary. Long paths from other latches or from elements like flip-flops or domino gates that impose hard edges may require time borrowing and thus dictate departure times. For the purpose of design, it suffices to nominally position latches at the beginning of the half-cycle; in the event of clock skew, data will automatically adjust using time borrowing to allow the system to operate at its minimum cycle time.

4.1.2   Static-to-Domino Interface

When nonmonotonic static signals are inputs to domino gates, they must be latched so that they will not change while the domino gate is in evaluation. The interface also imposes a hard edge because the data must set up at the domino input before the earliest the domino gate might begin evaluation, but may not propagate through the domino gate until the latest the gate could begin evaluation. Therefore, clock skew must be budgeted at the static-to-domino interface. This skew budget can be minimized by keeping the path in a local clock domain; Section 6.3 computes how much skew must be budgeted. Note that, strictly speaking, only falling inputs to domino gates must set up before evaluation. If your timing analyzer separately tracks arrival times for rising and falling edges, you can relax the constraints on the rising edge.

The latching technique at the interface depends on whether transparent or pulsed latches are used. In systems using transparent latches, static logic from one half-cycle can directly interface to dynamic logic at the start of the next half-cycle. The static outputs will not change while the domino is in evaluation. In systems with pulsed latches, however, the pulsed latch output may change while ϕ_1 domino gates are evaluating. Therefore, a modified pulsed latch must be used at the interface to pro-

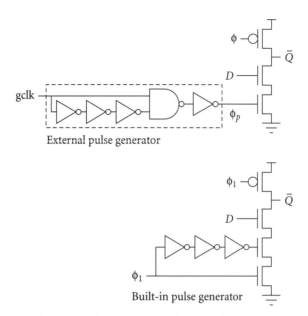

Figure 4.6 Pulsed domino latches with external and built-in pulse generators

duce monotonic outputs. This is called a "pulsed domino latch" [43] and is shown in Figure 4.6.

The pulsed domino latch essentially consists of a domino gate with a pulsed evaluation clock. The pulse may either be generated externally or produced by two series evaluation transistors as shown in the figure. The former scheme yields a faster latch because fewer series transistors are necessary, but requires longer pulses, as was discussed for ordinary pulsed latches in Section 2.3.2. Logic functions may be built into the pulsed domino latch.

The output of a static pulsed latch may be connected through static logic to ϕ_2 or ϕ_3 domino gates, so long as the static result settles before the domino enters evaluation. Master-slave flip-flops can be interfaced the same way. Neither static pulsed latches nor flip-flops directly interlace to ϕ_1 domino gates because the output is not monotonic during ϕ_1.

4.1.3 Domino-to-Static Interface

While a signal propagates through a skew-tolerant domino path, latches are unnecessary. However, before a domino output is sent into a block of static logic, it must be latched so that the result is not lost when the domino gate precharges. We will use a special latch at this interface that takes advantage of the monotonic nature of the domino outputs to improve performance.

Figure 4.7 shows the interface from domino to static logic. The dynamic gate drives a special latch using a single clocked NMOS transistor. This latch is called an N-C²MOS stage by Yuan and Svensson [96]; we will sometimes abbreviate it as an N-latch. When the dynamic gate evaluates, its falling transition propagates very quickly through the single PMOS transistor in the N-C²MOS latch. When the dynamic gate precharges and its output rises, the latch turns opaque, holding the value until the next time the clock rises. A weak keeper improves noise immunity when the clock is high. It is important to minimize the skew between the dynamic gate and N-C²MOS latch so that precharge cannot ripple through the latch. This is easy to do by locating the two cells adjacent to one another sharing the same clock wire. In Section 4.1.4, we will avoid this race entirely by using a latch clock that falls before the dynamic gate begins precharge. The only overhead at the interface from dynamic to static logic is the latch propagation delay.

The output Q of the circuit will always fall when the clock rises, then may rise depending on the input D. This results in glitches propagating through the static logic when D is held at a logic for multiple cycles; the glitches lead to excess power dissipation. When dual-rail domino signals are available, an SR latch can be used at the domino-to-static interface, as shown in Figure 4.8. The SR latch avoids glitches when the domino inputs do not change, but is slower because results must propagate through two NAND gates.

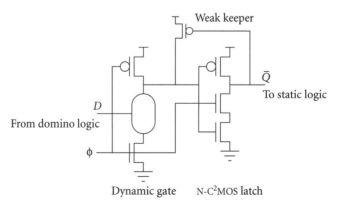

Figure 4.7   Domino-to-static interface

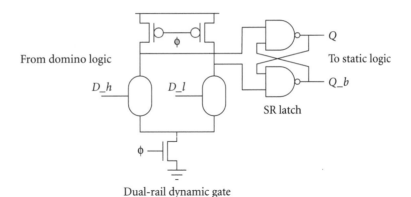

Figure 4.8   Glitch-free but slower domino-to-static interface

A regular transparent latch also can be used at the domino-to-static interface, but is slower than the N-C²MOS latch and has the same glitch problems.

4.1.4   Timing Types

The rules for connecting domino and static gates are complex enough that it is worthwhile systematically defining and checking the legal connectivity. To do this, we can generalize the two-phase clocking discipline rules of Noice [61] to handle four-phase skew-tolerant domino. Each signal name is assigned a suffix describing its timing. Proper connections can be verified by examining the suffixes. We first review the classical definition of timing types in the context of two-phase nonoverlapping clocks. Most systems use 50% duty cycle clocks, so we describe how timing types apply to such systems at the expense of checking min-delay. We then generalize timing types to four-phase skew-tolerant domino, including systems that mix domino, transparent latches, and pulsed latches. Timing types also include information about monotonicity and polarity to describe domino and dual-rail domino logic.

Two-Phase Nonoverlapping Clocks

Systems constructed from two-phase nonoverlapping clocks ϕ_1 and ϕ_2, like the one we examined in Figure 2.7, have the pleasant property that as long as simple topological rules are obeyed, the system will have no setup or hold time problems if run slowly enough with sufficient nonoverlap, regardless of clock skew [29]. They are particularly popular in student projects because no timing analysis is necessary. Timing types are used to specify the topological rules required for correct operation and allow automatic checking of legal connectivity. In later sections, we will extend timing types to work with practical systems that do not have nonoverlapping clocks. The extension comes at the expense of checking min-delay violations.

Each signal is given a suffix indicating the timing type and phase. The suffixes are _s1, _s2, _v1, _v2, _q1, and _q2. _s indicates that a signal is stable during a particular phase; that is, that the signal settles before the rising edge of the phase and does not change until after the falling edge. _v means that the signal is valid for sampling during a phase; it is stable for some setup and hold time around the falling edge of the phase. _q indicates a qualified clock, a glitch-free clock signal that may only rise on certain cycles. These timing types denote which clock edge controls the stability regions of the signals, that is, when the circuit is operated at slow speed, after which edge does the signal settle. Figure 4.9 shows examples of stable and valid signals. Stable is a stronger condition than valid; any stable signal can be used where a valid signal is required.

From the previous definitions, we see that clocks are neither valid nor stable. They establish the time and sequence references for data signals and are never latched by other clocks. However, it is sometimes useful to provide a clock that only pulses on certain cycles. _q indicates that the signal is such a qualified clock, a clock gated by some control so that it may

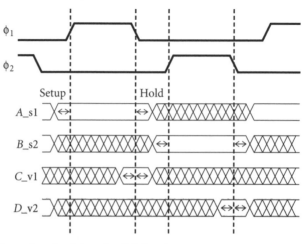

Figure 4.9   Examples of stable and valid signals

not rise during certain cycles. Clock qualification is discussed further in Section 4.1.5. Qualified clocks are interchangeable with normal clocks for the purpose of analysis.

The inputs to latches must be valid for a setup and hold time around the sampling edge of the latch clock. For the purpose of verifying correct operation with timing types, it is helpful to imagine operating the system at low frequency so all latch inputs arrive before the rising edge of the clock and no time borrowing is necessary. Thus, a latch output will settle sometime after the rising edge of the latch clock and will not change again until the following rising edge of the latch clock; hence it is stable throughout the other phase. If the system operates correctly at low speed, we can then increase the clock frequency, borrowing time until setup times no longer are met.

In summary, a ϕ_1 latch requires a _v1 or _s1 input and produces a _s2 output. A ϕ_2 latch requires a _v2 or _s2 input and produces a _s1 output. Combinational logic does not change timing types because the system can be operated slowly enough that data is still valid or stable in the specified phase. Figure 4.10 illustrates a general two-phase system.

Valid signals are produced by domino gates, as shown in Figure 4.11. The outputs settle sometime after the rising edge of the clock and do not

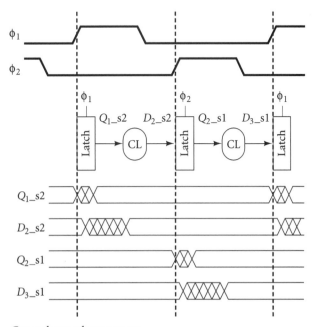

Figure 4.10   General two-phase system

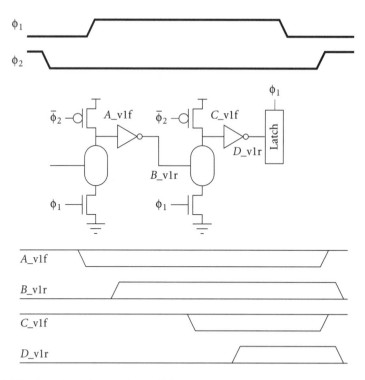

Figure 4.11   Domino gates produce valid outputs

precharge until the rising edge of the next clock, so they are valid for sampling around the falling edge of the clock. Using different precharge and evaluation clocks avoids any races between precharge and sampling. We also tag domino signals as monotonically rising (r) or falling (f). Domino inputs must be either stable or valid and monotonically rising during the phase the gate evaluates. The output of the dynamic gate is monotonically falling, and the output of the inverting static gate is monotonically rising. In such a traditional domino clocking scheme, the nonoverlap also appears as sequencing overhead.

As long as systems using two-phase nonoverlapping clocks have inputs to domino and latches using the proper timing types summarized in Table 4.1, the systems will always function at some speed. Setup time violations caused by long paths or excessive skew are solved by increasing the clock period. Hold time violations caused by short paths or excessive skew are solved by increasing the nonoverlap between phases.

Most two-phase systems use 50% duty cycle clocks rather than non-overlapping clocks. Timing types are still useful to describe legal connectivity, but clock skew can lead to hold time failures that cannot be

Table 4.1   Two-phase clocked element timing rules

Element type	Clock	Input	Output
Dynamic	ϕ_1, _q1	_s1, _v1r	_v1f
	ϕ_2, _q2	_s2, _v2r	_v2f
Transparent latch	ϕ_1, _q1	_s1, _v1	_s2
	ϕ_2, _q2	_s2, _v2	_s1

fixed by slowing the clock. Therefore, such systems must be checked for min-delay. In essence, the definitions of _v and _s must change to reflect the fact that the user can no longer control how long a signal will remain constant after the falling edge of a sampling clock. Also, since the two clocks are now complementary, domino gates use the same clock for evaluation and precharge. This leads to another hold time race as domino gates precharge at the same time as the latch samples. Timing types are still useful to indicate legal inputs to dynamic gates and transparent latches, but no longer guarantee immunity to min-delay problems.

Four-Phase Skew-Tolerant Domino

We can generalize the idea of timing types to four-phase skew-tolerant domino. Again, we will construct timing rules assuming that duty cycles can be adjusted to eliminate min-delay problems. Specifically, to avoid min-delay problems, each phase overlaps the next, but nonadjacent phases must not overlap, as shown in Figure 4.12. For example, ϕ_1 and ϕ_3 are nonoverlapping. In Section 4.1.6 we will consider the min-delay races that must be checked when the nonadjacent phases may overlap. We also use timing types to describe the interface of four-phase skew-tolerant domino with transparent latches, pulsed latches, and N-C²MOS latches.

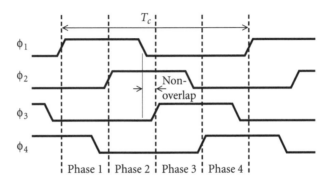

Figure 4.12   Ideal nonoverlapping clock waveforms

GUIDELINE 1   Each signal name must have a suffix that describes the timing, phase, monotonicity, and polarity.

The timing is s, v, or q and the phase is 1, 2, 3, 4, 12, 23, 34, or 41. This is similar to two-phase timing types, but extends the definitions to describe signals that are stable through more than one phase. The monotonicity may be r for monotonically rising, f for monotonically falling, or omitted if the signal is not monotonic during the phase. These suffixes are primarily applicable to domino circuits and skewed gates. Polarity may be any one of (blank), b, h, or lb indicates a complementary signal. h and l are used for dual-rail domino signals; when h is asserted, the result is a 1, while when l is asserted, the result is a 0. When neither is asserted, the result is not yet known, and when both are asserted, your circuit is in trouble. The signal is asserted when it is 1 for monotonically rising signals (r) and 0 for monotonically falling signals (f). Therefore, dual-rail dynamic gates produce fh and fl signals, while the subsequent dual-rail inverting static gates produce rh and rl signals. The suffix is written in the form

signalname_TP[M][Pol]

where T is the timing type, P is the phase, M is the monotonicity (if applicable), and Pol is the polarity (if applicable). For example, a sample path following these conventions is shown in Figure 4.13.

In addition to checking for correct timing and phase, we use timing types to verify monotonicity for domino gates.

Note that, unlike Figure 4.11, we now use the same clock for precharge and evaluation of dynamic gates. Therefore, the definition of a valid signal changes: a valid signal settles before the falling edge of the clock and does not change until shortly after the falling edge of the clock. This is exactly the same timing rule as a qualified clock, so _v and _q signals are now in principle interchangeable. Nevertheless, we are much more con-

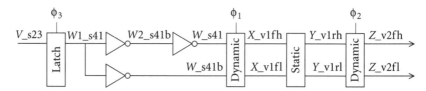

Figure 4.13   Path illustrating timing types and static-to-domino interface

cerned about controlling skew on clocks, so we reserve the _q timing type for clock signals and continue to use _v for dynamic outputs with the understanding that the length of the validity period is not as great as it was in a classical two-phase system. In particular, a _v1 signal is not a safe input to a ϕ_1 static latch because the dynamic gate may precharge at the same time the latch samples. For example, Figure 4.14 illustrates how latch B's output might incorrectly fall when dynamic gate A precharges if there is skew between the clocks of the two elements.

DEFINITION 2   The _v inputs to a dynamic gate must be monotonically rising (r). The output of a dynamic gate is monotonically falling (f).

This definition formalizes the requirement of monotonicity. _s inputs to a dynamic gate stabilize before the gate begins evaluation, so they do not have to be monotonic. As an example, a ϕ_1 dynamic gate may take input X_v4r or Y_s1 or even Z_s1f but not W_v4f.

DEFINITION 3   Inverting static gates that accept exclusively monotonically rising inputs (r) produce monotonically falling outputs (f) and vice versa.

GUIDELINE 2   Static gates should be skewed HI for monotonically falling (f) inputs and LO for monotonically rising (r) inputs.

Skewed gates may use different P/N ratios to favor the critical transitions and improve speed. HI-skew gates with large P/N ratios should follow monotonically falling dynamic outputs. When a path built with static logic

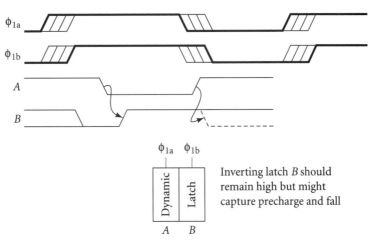

Figure 4.14   Potential race at interface of _v1 signal to ϕ_1 static latch

is monotonic, alternating HI- and LO-skew gates can be used for speed. A normal inverter has a P/N ratio of about 2/1. A HI-skew inverter may have a P/N ratio of 4/1, while a LO-skew inverter has a P/N ratio of 1/1 to favor the critical transitions.

GUIDELINE 3   _s and _v signals are the only legal data inputs to gates and latches. _q and ϕ are the only legal clock inputs.

This is identical to traditional two-phase conventions. Clocks and gates should not mix except at clock qualifiers (see Guideline 7).

GUIDELINE 4   The output timing type of a static gate is the intersection of the input types. If the intersection is empty, the gate is receiving an illegal set of inputs.

Remember that a _v signal can be sampled for a subset of the time that a _s signal of the same phase is stable. For example, a gate receiving _s12 and _v2 inputs produces a _v2 output. A gate receiving _s12 and _s34 inputs has no legal output timing type, so the inputs are incompatible.[1]

GUIDELINE 5   Table 4.2 summarizes timing rules for the most common elements in a system mixing skew-tolerant domino and transparent latches.

Table 4.2   Simplified clocked element timing guidelines

Element type	Clock	Input	Output
Dynamic	ϕ_1, _q1	_s1, _v4r (first), _v1r (others)	_v1f
	ϕ_2, _q2	_s2, _v1r (first), _v2r (others)	_v2f
	ϕ_3, _q3	_s3, _v2r (first), _v3r (others)	_v3f
	ϕ_4, _q4	_s4, _v3r (first), _v4r (others)	_v4f
Transparent latch	ϕ_1, _q1	_s1	_s23
	ϕ_3, _q3	_s3	_s41
N-C^{2}MOS latch	ϕ_1, _q1	_v2f	_s3r
	ϕ_3, _q3	_v4f	_s1r

---

1. This guideline has an obscure exception. A static gate may receive both _s2 and _v1f inputs. Its output is a special type (let us say _v1rs2) only suitable as an input to a ϕ_2 dynamic gate. This would occur when an inverting static gate receives inputs from a ϕ_1 static latch and a ϕ_1 dynamic gate in order to drive a ϕ_2 dynamic gate. Similar combinations are legal inputs to other phases of domino gates.

Clocked elements set the output timing type and require inputs which are valid when they may be sampled. The types depend on the clock phase used by the clocked element. See Table 4.3 (Guideline 6) for a complete list of timing rules covering more special cases.

The output of a dynamic gate is valid and monotonically falling during the phase the gate operates, just as we have seen for two-phase systems. The input can come from static or domino logic. Static inputs must be stable _s while the gate is in evaluation to avoid glitches. Inputs from domino logic are monotonic rising and thus only must be valid _v. The key difference between conventional timing types and skew-tolerant timing types is that valid inputs to the first dynamic gate in each phase come from the previous phase, while inputs to later dynamic gates come from the current phase. Technically, different series stacks may receive _v inputs from different phases.

N-latches are used at the interface of domino to static logic. Although transparent latches could also be used, they are slower and present more clock load, so they are not recommended in this methodology. Notice that N-latches use a clock from one phase earlier than the dynamic gate they are latching to avoid race conditions by which that dynamic gate may precharge before the latch becomes opaque. Because of the single PMOS pullup in the N-latch, dynamic gate output A evaluating late can borrow time through the N-latch even after the latch clock falls, as shown in Figure 4.15.

The N-latch output B settles after the rising clock edge evaluating the preceding dynamic gate and remains stable until the next rising edge of the latch clock, so it is stable for one phase (Phase 3, in this example). Because the interface between dynamic and static logic normally occurs at half-cycle boundaries, the ϕ_2 and ϕ_4 N-latches are rarely used.

The output of a transparent latch stabilizes after the latch becomes transparent and remains stable until the latch becomes transparent again; for example, the output of a ϕ_1 transparent latch is _s23. Signals stable for multiple phases are legal as inputs to elements requiring stability in either phase. For example, a _s23 signal is a legal input to a ϕ_3 transparent latch that requires _s3. While transparent latches technically can accept certain _v inputs as we saw with two-phase timing types, N-latches are preferred at this interface of domino to static so only the _s inputs are shown in the table for transparent latches.

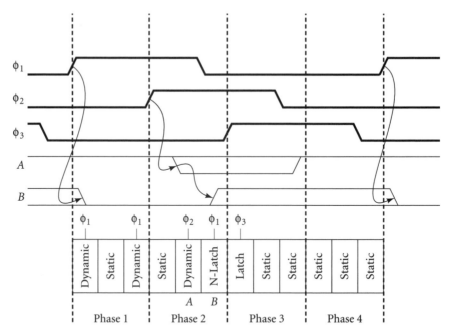

Figure 4.15 Domino-to-static interface using N-latch

Figure 4.16 illustrates legal connections between static and domino gates. The top row of gates is a pipeline of static logic, while the bottom row is domino. Dynamic outputs must pass through N-C^{2}MOS latches before driving static logic. Static signals can directly drive appropriate domino gates.

Figure 4.17 illustrates illegal connections between static and domino gates. Connection A is bad because the dynamic output will precharge while the ϕ_3 latch is transparent, allowing incorrect data to propagate into the static logic. An N-latch should have been used at the interface. Connection B is bad because the static path skips over a latch. Connection C is bad because the ϕ_4 domino gate receives both _v3 and _v4 inputs. By the time the _v4 input arrives, the _v3 result may have precharged. Each of these illegal connections violates the possible inputs of Table 4.2, so an automatic tool could flag these errors.

Now that we have seen how timing types work, we can expand them to handle pulsed latches and uncommon phases of latches. The complete guidelines are summarized in Table 4.3.

GUIDELINE 6 Inputs and outputs of clocked elements should match the timing types defined in Table 4.3.

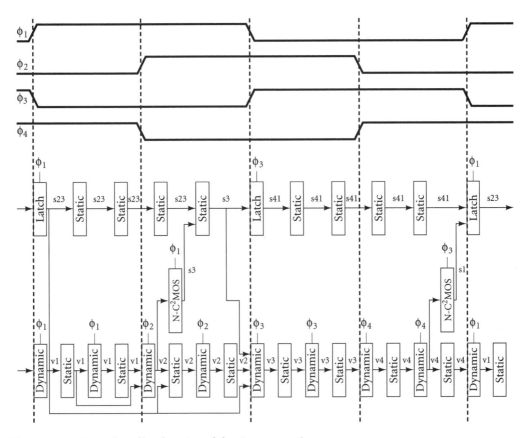

Figure 4.16 Examples of legal static and domino connections

Table 4.3 Complete clocked element timing guidelines

Element type	Clock	Input	Output
Domino	ϕ_1, _q1	_s1, _v4r (first), _v1r (others)	_v1f
	ϕ_2, _q2	_s2, _v1r (first), _v2r (others)	_v2f
	ϕ_3, _q3	_s3, _v2r (first), _v3r (others)	_v3f
	ϕ_4, _q4	_s4, _v3r (first), _v4r (others)	_v4f
Transparent latch	ϕ_1, _q1	_s1	_s23
	ϕ_2, _q2 (rare)	_s2	_s34
	ϕ_3, _q3	_s3	_s41
	ϕ_4, _q4 (rare)	_s4	_s12
Pulsed latch	ϕ_1, _q1	_s1, (_s2, _s3,) _s4, _v4	_s23
Pulsed domino latch	ϕ_1, _q1	_s1, (_s2, _s3,) _s4, _v4	_v1r
N-C^{2}MOS latch	ϕ_1, _q1	_v2f	_s3r
	ϕ_2, _q2 (rare)	_v3f	_s4r
	ϕ_3, _q3	_v4f	_s1r
	ϕ_4, _q4 (rare)	_v1f	_s2r

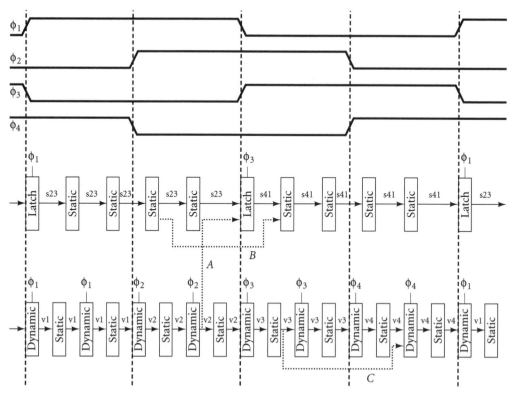

Figure 4.17   Examples of illegal static and domino connections

ϕ_2 and ϕ_4 transparent latches are not normally used. They occasionally are useful, however, in short paths to avoid min-delay problems as we shall see in Section 4.1.6. ϕ_2 and ϕ_4 N-latches are also rare, but may be used to interface from the middle of a half-cycle of domino logic back to static logic. These rare latches have timing analogous to their more common counterparts.

Pulsed latches are controlled by a brief pulse derived from the rising edge of ϕ_1. They accept any signal that will not change on or immediately after this edge (i.e., _s1, _s4, and _v4). The output has the same stable time as the output of a ϕ_1 transparent latch because it stabilizes after the rising edge of ϕ_1 and does not change until after the next rising edge of ϕ_1. Unfortunately, we see that the output of a pulsed latch is _s23, but neither _s2 nor _s3 signals are safe inputs to pulsed latches, so it is unsafe for one pulsed latch to drive another. This matches our understanding that pulsed latches cannot be cascaded directly without logic between them because of hold time problems. In order to build systems with pulsed latches, we relax the timing rules to permit _s2 and _s3 inputs to pulsed latches, then check for min-delay on such inputs. Such checks are discussed further in Section 4.1.6.

Pulsed domino latches have the same input restrictions as pulsed latches, but produce a _v1r output suitable for domino logic because their outputs become valid after the rising edge of ϕ_1 and remain valid until the falling edge of ϕ_1 when the gate precharges.

4.1.5 Qualified Clocks

Qualified clocks are used to save power by disabling units or to build combination multiplexer-latches in which only one of several parallel latches is activated each cycle. Qualification must be done properly to avoid glitches.

GUIDELINE 7 Qualified clocks are produced by ANDing ϕ_i with a _s(i) signal in the clock buffer.

To avoid problems with clock skew, it is best to qualify the clock with a signal that will remain stable long after the falling edge of the clock. For example, Figure 4.18 shows two ways to generate a _q1 clock. The _s qualification signal must set up before ϕ_1 rises and should not change until after ϕ_1 falls. In the left circuit, we AND ϕ_1 with a _s41 signal. If there is clock skew, the _s41 signal may change before ϕ_1 falls, allowing the _q1 clock to glitch. Glitching clocks are very bad, so the right circuit, in which we AND ϕ_1 with a _s12 signal, is much preferred. This problem is analogous to min-delay. Like min-delay, it could also be solved by delaying the _s41 signal so that it would not arrive at the AND gate before the falling edge of ϕ_1. However, clock qualification signals are often critical, so it is unwise to delay them unnecessarily. The race could also be solved by making the skew between the ϕ_1 and ϕ_3 clocks in the left circuit small.

4.1.6 Min-Delay Checks

We have noted that two-phase systems usually use complementary clocks rather than nonoverlapping clocks and thus lose their strict safety properties, requiring explicit checks for min-delay violations. Similarly, the four-phase timing types of Section 4.1.4 use nonoverlapping ϕ_1 and ϕ_3 to achieve safety, but real systems typically would use 50% duty cycle clocks.

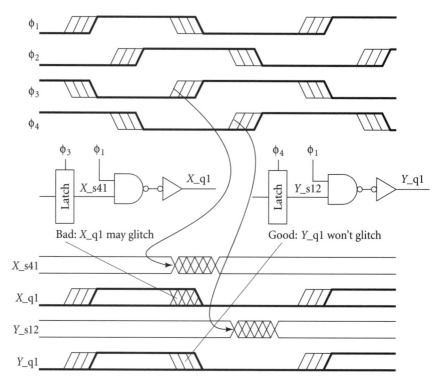

Figure 4.18   Good and bad methods of qualifying a clock

In this section, we describe where min-delay risks arise with 50% duty cycle clocks. We also examine the min-delay problems caused by pulsed latches.

Min-delay is a serious problem because unlike setup time violations, hold time violations cannot be fixed by adjusting the clock frequency. Instead, the designer must conservatively guarantee adequate delay through logic between clocked elements. Min-delay problems should be checked at the interfaces listed in Table 4.4. The top half of the table lists common cases encountered in typical designs. The bottom half of the table lists combinations, which while technically legal according to Table 4.2, would not occur in normal use because ϕ_2 and ϕ_4 transparent latches and N-latches are seldom used.

Min-delay problems can be solved in two ways. One approach is to add explicit delay to the data. For example, a buffer made from two long-channel inverters is a popular delay element. Another is to add a latch between the elements controlled by an intervening phase. Both approaches prevent races by slowing the data until the hold time of the second ele-

Table 4.4   Interfaces prone to min-delay problems

Source element	Source phase	Destination element	Destination phase
Transparent latch or N-latch or pulsed latch	ϕ_1	Transparent latch or dynamic gate	ϕ_3
Transparent latch or N-latch	ϕ_3	Transparent latch or dynamic gate	ϕ_1
Transparent latch or N-latch or pulsed latch	ϕ_1	Pulsed latch or pulsed domino latch	ϕ_1
Transparent latch or N-latch	ϕ_2	Transparent latch or dynamic gate	ϕ_4
Transparent latch or N-latch	ϕ_4	Transparent latch or dynamic gate	ϕ_2

ment is satisfied. Examples of these solutions are shown in Figure 4.19. In path (a) there is no logic between latches. If ϕ_1 and ϕ_3 are skewed as shown, data may depart the ϕ_1 latch when it becomes transparent, then race through the ϕ_3 latch before it becomes opaque. Path (b) solves this problem by adding logic delay δ_{logic}, as was discussed in Section 2.2.4. Path (c) solves the problem by adding a ϕ_2 latch. If the minimum required delay is large, the latch may occupy less area than a string of delay elements.

Min-delay problems can actually occur at any interface, not just those listed in Table 4.4. For example, if clock skew were greater than a quarter cycle, min-delay problems could occur between ϕ_1 and ϕ_2 transparent latches. Because it is very difficult to design systems when the clock skew exceeds a quarter cycle, we will avoid such problems by requiring that the clock have less than a quarter cycle of skew between communicating elements.

Depending on the clock generation method, a few other connections may incur races. It is possible to construct clock generators with adjustable delays so that as the frequency reduces, the delay between each phase remains T_c/N. However, as we will see in Section 5.2.1, it may be more convenient to produce ϕ_2 and ϕ_4 by delaying ϕ_1 and ϕ_3, respectively, by a fixed amount. Such clock generators introduce the possibility of races that are frequency independent because the delay between phases is fixed.

One such risky connection is a ϕ_1 pulsed latch feeding a ϕ_2 domino gate. There is a max-delay condition that data must set up at the input of the domino gate before the gate enters evaluation. Clock skew between ϕ_1 and ϕ_2 reduces the nominally available quarter cycle. Since the delay from

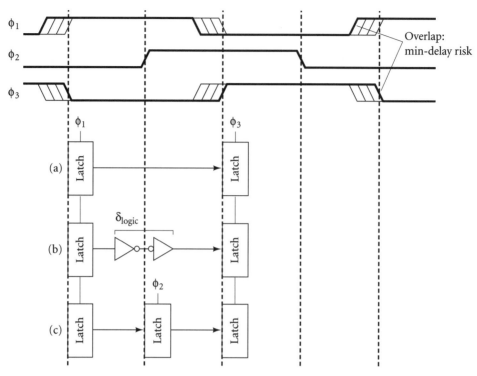

Figure 4.19   Solution to min-delay problem between ϕ_1 and ϕ_3 transparent latches: min-delay risk (a), extra gates (b), and extra latch (c)

ϕ_1 to ϕ_2 is constant, if domino input does not set up in time, the circuit will fail at any clock frequency. The same problem occurs at the interface of a ϕ_1 transparent latch to ϕ_2 domino and of a ϕ_3 transparent latch to ϕ_4 domino.

Another min-delay problem occurs between ϕ_2 transparent latches and ϕ_1 pulsed latches or pulsed domino latches. Again, if the delay between phases is independent of frequency, hold time violations cannot be fixed by adjusting the clock frequency.

4.2   Clocked Element Design

This section offers guidelines on the design of fast clocked elements. Remember that static CMOS logic uses either transparent latches or pulsed latches. Domino logic uses no latches at all, except at the interface back to static logic, where N-latches should be used. We postpone discussion of supporting scan in clocked elements until Section 4.3.

Critical paths should be entirely domino wherever possible because we must budget skew and latch propagation delay when making a transition from static back into domino logic; moreover, time borrowing is not possible through the interface. Because most functions are nonmonotonic, this frequently dictates using dual-rail domino. In certain cases, dual-rail domino costs too much area, routing, or power. For high-speed systems, going entirely static may be faster than mixing domino and static and paying the interface overhead. If the overhead is acceptable because skew is tightly controlled, try combining as much of the nonmonotonic logic into static gates at the end of the block, as discussed in Section 3.2.1.

4.2.1   Latch Design

The fastest latches are simply transmission gates. To avoid the noise problems described in Section 2.3, the gates should be preceded and followed by static CMOS gates. These gates may perform logic rather than merely being buffers, so the latch presents very little timing overhead. Pulsed latches can be produced by simply pulsing the transmission gate control.

GUIDELINE 8   Use a transmission gate as a latch receiving input from a static logic block. Use a full keeper on the output for static operation.

The transmission gate latch is very fast and compact and is used in the DEC Alpha 21164 methodology. The output is a dynamic node and must obey dynamic node noise rules. Therefore, it should drive a static gate not far from the output. A ϕ_1 static latch is shown in Figure 4.20.

GUIDELINE 9   Use a static gate before and after each static latch.

This gate is conventionally an inverter, but may be changed to any other static gate. The static gate after the latch should have no more than two inputs to avoid charge-sharing problems when the channel is not conducting. There should be little routing between the input gate, transmission gate, and output gate to minimize power supply noise problems and coupling onto the dynamic node. For synthesized blocks, it may be best to create versions of the latches incorporating gates into the input

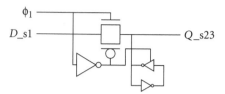

Figure 4.20 ϕ_1 static latch

and output as a single cell because synthesis tools have difficulty estimating the delay of an unbuffered transmission gate and because place-and-route tools may produce unpredictable routing.

GUIDELINE 10   Generally use only pulsed latches or ϕ_1 and ϕ_3 transparent latches.

We select two phases to be the primary static latch clocks to resemble traditional two-phase design. The ϕ_2 and ϕ_4 clocks would be confusing if generally used, so they are restricted to use to solve min-delay problems in short paths.

GUIDELINE 11   Use an N-C²MOS latch on the output of domino logic driving static gates as shown in Figure 4.7. Use a full keeper on the output for static operation.

Again, the output is a dynamic node and must obey dynamic node noise rules. The N-latch is selected because it is faster and smaller than a tristate latch and doesn't have the charge-sharing problems seen if a domino gate drove a transmission gate latch.

GUIDELINE 12   The domino gate driving an N-latch should be located adjacent to the latch and should share the same clock wire and V_{DD} as the latch.

The N-latch has very little noise margin for noise on the positive supply. This noise can be minimized by keeping the latch adjacent to the domino output, thereby preventing significant noise coupling or V_{DD} drop. The latch is also sensitive to clock skew because if it closed too late, it could capture the precharge edge of the domino gate. Therefore, the same clock wire should be used to minimize skew.

4.2.2   Domino Gate Design

The guidelines in this section cover keepers, charge-sharing noise, and unfooted domino gates.

GUIDELINE 13   All dynamic gates must include a keeper.

> In Section 4.3 we will see that the clock is stopped low during test, so a keeper is necessary to fight leakage on ϕ_3 and ϕ_4 dynamic gates. It is also necessary on all gates to achieve reasonable noise immunity. Breaking this guideline requires extremely careful attention to dynamic gate input noise margins.

GUIDELINE 14   The first dynamic gate of Phase 3 must include a full keeper.

> As discussed in Section 3.2.3, this is necessary to prevent the outputs of the first Phase 3 gates from floating when the clock is stopped low and the Phase 2 gates precharge. Note that because the first dynamic gate of Phase 1 does not include a full keeper, the clock should not be stopped high long enough for the output to be corrupted by subthreshold leakage. Of course, this guideline is an artifact of the methodology; an alternative methodology that stopped the clock high or allowed clock stopping both high and low would require the full keeper on Phase 1. In Section 4.3.2 we will see that the last dynamic gate of Phase 4 may also need a full keeper to support scan.

GUIDELINE 15   Use secondary precharge devices on internal nodes to control charge-sharing noise.

> The exact number of secondary precharge devices required depends on the noise budget. A reasonable rule of thumb is to precharge every other internal node (starting from the top), plus any internal node with a large amount of diffusion, as discussed in Section 3.2.4.

GUIDELINE 16   The output of a dynamic gate must drive the gate, not source/drain input of the subsequent gate.

> The result of a dynamic gate is stored on the capacitance of the output node, so this guideline prevents charge-sharing problems. An important

implication is that dynamic gates cannot drive transmission gate multiplexer data inputs, although they could drive tristate-based multiplexers.

GUIDELINE 17   Use footed dynamic gates exclusively.

This guideline is in place to avoid excess power consumption, which may occur when the pulldown transistors are all on while the gate is still precharging. It may be waived on the first ϕ_2 and ϕ_4 gates of each cycle so long as the inputs of the gates come from ϕ_1 or ϕ_3 domino logic that does not produce a rising output until the ϕ_2 or ϕ_4 gates have entered evaluation. Aggressive designers may waive the guideline on other dynamic gates if power consumption is tolerable.

4.2.3   Special Structures

In a real system, skew-tolerant domino circuits must interface to special structures such as memories, register files, and programmable logic arrays (PLAs). Precharged structures like register files are indistinguishable in timing from ordinary domino gates. Indeed, standard six-transistor register cells can produce dual-rail outputs suitable for immediate consumption by dual-rail domino gates.

Certain very useful dynamic structures such as wide comparators and dynamic PLAs are inherently nonmonotonic and are conventionally built for high performance using self-timed clocks to signal completion. The problem is that these structures are most efficiently implemented with cascaded wide dynamic gates because the delay of a dynamic NOR structure is only a weak function of the number of inputs. Generally, dynamic gates cannot be directly cascaded. However, if the second dynamic gate waits to evaluate until the first gate has completed evaluation, the inputs to the second gate will be stable and the circuit will compute correctly. The challenge is creating a suitable delay between gates. If the delay is too long, time is wasted. If the delay is too short, the second gate may obtain the wrong result.

A common solution is to locally create a self-timed clock by sensing the completion of a model of the first dynamic gate. For example, Figure 4.21 shows a dynamic NOR-NOR PLA integrated into a skew-tolerant pipeline. The AND plane is illustrated evaluating during ϕ_2, and adjacent logic can evaluate in the same or nearby phases. *andclk* is nominally in phase with ϕ_2, but has a delayed falling edge to avoid a precharge

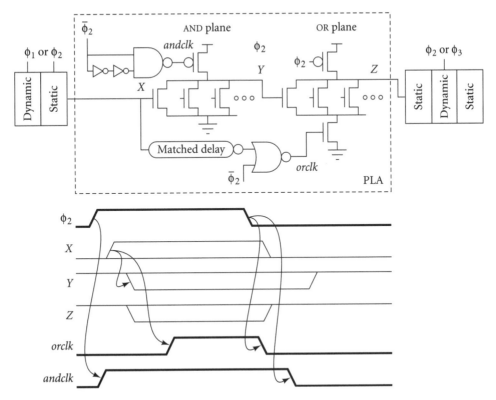

Figure 4.21 Domino/PLA interface

race with the OR plane. The latest input X to the AND plane is used by a dummy row to produce a self-timed clock *orclk* for the OR plane that rises after AND plane output Y has settled. Notice how the falling edge of *orclk* is not delayed so that when Y precharges high the OR plane will not be corrupted. The output Z of the OR plane is then indistinguishable from any other dynamic output and can be used in subsequent skew-tolerant domino logic.

4.3 Testability

As integrated circuits use ever more transistors and overlay the transistors with an increasing number of metal layers, debug and functional test become more difficult. Packaging advances such as flip-chip technology make physical access to circuit nodes harder. Hence, engineers employ design for testability methods, trading area and even some amount of performance to facilitate test. The most important testability technique is scan, in which memory elements are made externally observable and controllable through a scan chain [66]. Scan generally involves modifying flip-flops or latches to add scan signals.

Because scan provides no direct value to most customers, it should impact a design as little as possible. A good scan technique has

- minimal performance impact

- minimal area increase

- minimal design time increase

- no timing-critical scan signals

- little or no clock gating

- minimal tester time

The area criterion implies that scan should add little extra cell area and also few extra wires. The timing-critical scan signal criterion is important because scan should not introduce critical paths or require analysis and timing verification of the scan logic. Clock gating is costly because it increases clock skew and may increase the setup on already critical clock enable signals such as global stall requests.

We will assume that scan is performed by stopping the global clock low (i.e., ϕ_1 and ϕ_2 low and ϕ_3 and ϕ_4 high), then toggling scan control signals to read out the current contents of memory elements and write in new contents. We will first review scan of transparent and pulsed latches, then extend the method to scan skew-tolerant domino gates in a very similar fashion.

4.3.1 Static Logic

Systems built from transparent latches or pulsed latches can be made observable and controllable by adding scan circuitry to every cycle of logic. Figure 4.22 shows a scannable latch. Normal latch operation involves input D, output Q, and clock ϕ. When the clock is stopped low, the latch is opaque. The circuits shown in the dashed boxes are added to the basic latch for scan. The contents of latch can be scanned out to *SDO* (scan data out) and loaded from *SDI* (scan data in) by toggling the scan clocks *SCA* and *SCB*. While it is possible to use a single scan clock, the

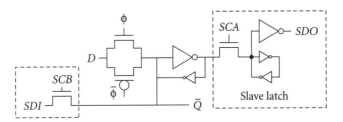

Figure 4.22 Scannable latch

two-phase nonoverlapping scan clocks shown are more robust and simplify clock routing. The small inverters represent weak feedback devices; they must be ratioed to allow proper writing of the cell. Note that this means the gate driving the data input D must be strong enough to overpower the feedback inverter. Although a tristate feedback gate may be used instead, it must still be weak enough to be overpowered by SDI during scan.

We assume that scan takes place while the clock is stopped low. Therefore, transparent latch systems make the first half-cycle latch scannable, and pulsed latch systems make the pulsed latch scannable. The procedure for scan is

1 Stop gclk low.

2 Toggle SCA and SCB to march data through the scan chain.

3 Restart gclk.

GUIDELINE 18 Make all pulsed latches and ϕ_1 transparent latches scannable.

4.3.2 Domino Logic

Because skew-tolerant domino does not use latches, some other method must be used to observe and control one node in each cycle. Controlling a node requires cutting off the normal driver of the node and activating an access transistor. For example, latches are controlled during scan by being made opaque, then activating the SCB access transistor in Figure 4.22. A dynamic gate with a full keeper can be controlled in an analogous way by turning off both the evaluation and precharge transistors and turning on an access transistor, as shown in Figure 4.23. Notice that separate evaluation and precharge signals are necessary to turn off both devices so a

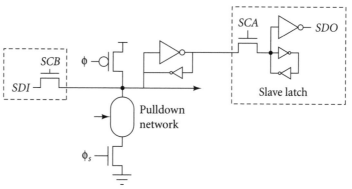

Figure 4.23   Scannable dynamic gate

gated clock ϕ_s is introduced. A slave latch connected to the full keeper provides observability without loading the critical path, just as it does on a static latch. Note that this is a ratioed circuit and the usual care must be taken that feedback inverters are sufficiently weak in all process corners to be overpowered. Also, note that only a PMOS keeper is required on the dynamic output node if *SCA* and *SCB* are toggled quickly enough that leakage is not a problem.

Which dynamic gate in a cycle should be scannable? The gate should be chosen so that during scan, the subsequent domino gate is precharging so that glitches will not contaminate later circuits. The gate should also be chosen so that when normal operation resumes, the output will hold the value loaded by scan until it is consumed.

Let us assume that scan is done while the global clock is stopped low, thus with the ϕ_1 and ϕ_2 domino gates in the first half-cycle precharging and the ϕ_3 and ϕ_4 gates in the second half-cycle evaluating. Then a convenient choice is to scan the last ϕ_4 domino gate in the cycle. This means that the last ϕ_4 domino gate must include a full keeper. Scan is done with the following procedure:

1 Stop gclk low.

2 Stop ϕ_s low.

3 Toggle *SCA* and *SCB* to march data through the scan chain.

4 Restart gclk.

5 Release ϕ_s once scannable gate begins precharge.

When gclk is stopped, the scannable gate will have evaluated to produce a result. Stopping ϕ_s low will turn off the evaluation transistor to the scannable gate, leaving the output on the dynamic node held only by the full keeper. Toggling *SCA* and *SCB* will advance the result down the scan chain and load a new value into the dynamic gate. When gclk restarts, it rises, allowing the gates in the first half-cycle to evaluate with the data stored on the scan node. Once the scannable gate begins precharging, ϕ_s can be released because the gate no longer needs to be cut off from its inputs.

Unfortunately, this scheme requires releasing ϕ_s in a fraction of a clock cycle. It would be undesirable to do this with a global control signal because it is difficult to get a global signal to all parts of the chip in a tightly controlled amount of time. It is better to use a small amount of logic in the local clock generator to automatically perform Steps 2 and 5. We will examine such a clock generator supporting four-phase skew-tolerant domino with clock enabling and scan in Section 5.2.3.

A potential difficulty with scanning dynamic gates is that it could double the size of a dynamic cell library if both scannable and normal dynamic gates are provided. A solution to this problem is to provide a special scan cell that "bolts on" to an ordinary dynamic gate. The scan cell adds a full keeper and scan circuitry to the ordinary gate's PMOS keeper, as shown in Figure 4.24. In ordinary usage, the two clock inputs of the dynamic gate are shorted to ϕ_4, while in a scannable gate ϕ_4 and ϕ_{4s} are connected.

GUIDELINE 19   Make the last domino gate of each cycle scannable with bolt-on scan logic.

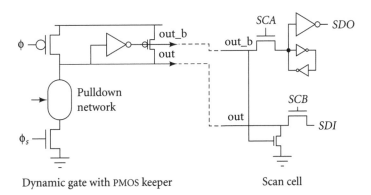

Dynamic gate with PMOS keeper   Scan cell

Figure 4.24   Dynamic gate and scan cell

A cycle may combine static and domino logic. As long as all first half-cycle latches and the last domino gate in each second half-cycle are scanned, the cycle is fully testable. Static and domino scan nodes are compatible and may be mixed in the same scan chain. Note that pulsed domino latches are treated as first half-cycle domino gates and are not scanned.

4.4 Historical Perspective

Circuit methodologies are seldom published, although employee turnover in Silicon Valley tends to carry the best techniques from one company to another in a clandestine fashion. The best-documented methodologies have been from IBM and DEC. DEC (now Compaq) has led the pack in circuit performance since the introduction of the Alpha 21064, so we will follow the evolution of their circuit methodology. In the next historical perspective, we will follow the DEC clocking methodology.

The Alpha 21064 [16, 17] arrived in 1993 at a blazing 200 MHz while other microprocessors were running 40–100 MHz. This speed corresponds to approximately 20 fanout-of-4 inverter delays per cycle in the 0.75-micron process. Earlier DEC chips had used a four-phase nonoverlapping clock system distributed on multiple clock wires to avoid risk of min-delay. To achieve acceptable skew at 200 MHz, the Alpha used a single global clock wire. The processor used true single-phase clock (TSPC) transparent latches and avoided min-delay problems through clock distribution and latch design. Logic could be built into the first stage of the TSPC latches.

The Alpha 21164 [5, 6, 27] arrived two years later at 300 MHz, using fewer than 18 FO4 inverter delays per cycle in the 0.5-micron process [30]. The designers found that an ordinary transparent latch was faster than the TSPC latch [26], especially because logic could be integrated into both the gate before and after the latch. Min-delay was avoided by requiring at least one minimum logic delay element between all latches and was verified by a simple CAD tool. The 21164 was also the first published processor to eliminate a latch from the middle of a cycle of domino logic by overlapping clocks.

Curiously, the Alpha 21264 [28] broke away from the skew-tolerant techniques advocated in this book. It shipped at 575 MHz in late 1998 in a

0.35-micron process. The processor uses an unusual edge-triggered flip-flop [28] instead of transparent latches. This apparently sacrifices skew tolerance and increases the latching overhead as a fraction of cycle time, although it does provide a sense amplifier for special low-swing differential signals. Min-delay risks are increased due to the more complex clocking scheme and are avoided by careful checking with a CAD tool.

4.5 Summary

This chapter has described a method for designing systems with transparent and pulsed latches and skew-tolerant domino. It uses a single globally distributed clock from which four local overlapping clock phases are derived. The methodology supports stopping the clock low for power savings and testability and describes a low-overhead scan technique compatible with both domino and static circuits. Timing types are used to verify proper connectivity among the clocked elements.

The methodology hides sequencing overhead everywhere except at the interface between static and domino logic. At the interface of domino-to-static logic, a latch is necessary to hold the result, adding propagation delay to the critical path. More importantly, at the interface of static-to-domino logic, clock skew must be budgeted so that inputs settle before the earliest the evaluation clock might rise, yet the domino gate may not begin evaluation until the latest time the clock might rise. This overhead makes it expensive to switch between static and domino logic. Designers who need domino logic to meet cycle time targets should therefore consider implementing their entire path in domino. Because single-rail domino cannot implement nonmonotonic functions, dual-rail domino is usually necessary. Therefore, we should expect to see more critical paths built entirely from dual-rail domino as sequencing overhead becomes a greater portion of the cycle time.

4.6 Exercises

[15] **4.1** Define time borrowing. Which of the following circuits permit time borrowing: transparent latches, pulsed latches, flip-flops, traditional domino circuits, skew-tolerant domino circuits?

[25]   **4.2**   Draw timing loop diagrams like those of Figure 4.4 for the path in Figure 4.25 assuming cycle times of 1.5, 1, and 0.8 units using the following combinational logic delays. Note which latches borrow time and if setup time violations occur.

(a) $\Delta1 = 0.5$; $\Delta2 = 0.95$; $\Delta3 = 1.45$; $\Delta4 = 0.3$; $\Delta5 = 0.2$

(b) $\Delta1 = 0.3$; $\Delta2 = 0.4$; $\Delta3 = 0.81$; $\Delta4 = 0.8$; $\Delta5 = 0.8$

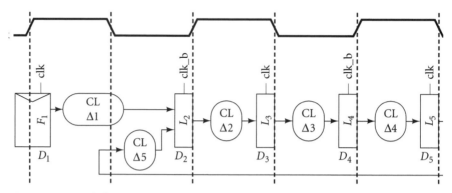

Figure 4.25   Path for Exercises 4.2 and 4.3

[15]   **4.3**   Label the timing types of each connection in the circuit in Figure 4.25 using the two-phase timing types defined in this chapter. Assume the flop is built from two back-to-back transparent latches, the first controlled by clk_b and the second controlled by clk.

[20]   **4.4**   Label the timing types of each connection in the circuit in Figure 4.26 using the two-phase timing types defined in this chapter.

[20]   **4.5**   Label the timing types of each connection in the circuit in Figure 4.27 using the four-phase domino timing types defined in this chapter.

[40]   **4.6**   CAD project: Construct a time_check tool. The tool should accept a netlist of four types of elements: S (static), D (dynamic), L (transparent latch), and N (N-C^{2}MOS latch). Each element has an output and one or more inputs. Nonstatic elements also have a clock input. The tool should identify the timing type of each net and report elements with illegal input timing types.

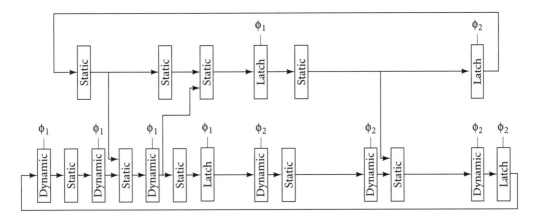

Figure 4.26 Path for Exercise 4.4

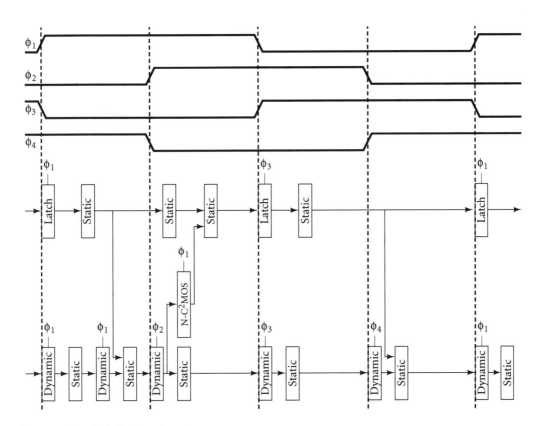

Figure 4.27 Path for Exercise 4.5

[15]   **4.7** Determine the minimum delay δ_{logic} between the circuits in Table 4.5 in terms of δ_{CQ}, δ_{CD}, t_{skew}, and the pulsed latch pulse width t_{pw}.

Table 4.5   Paths for Exercise 4.7

	Source element	Source phase	Destination element	Destination phase
(a)	Transparent latch	ϕ_1	Transparent latch	ϕ_3
(b)	Pulsed latch	ϕ_1	Transparent latch	ϕ_3
(c)	Transparent latch	ϕ_1	Pulsed latch	ϕ_1

5
Clocking

Clocking is a key challenge for high-speed circuit designers. Circuit designers specify a modest number of *logical* clocks that ideally arrive at all points on the chip at the same time. For example, flip-flop–based systems use a single logical clock, while skew-tolerant domino might use four logical clocks. Unfortunately, mismatched clock network paths and processing and environmental variations make it impossible for all clocks to arrive at exactly the same time, so the designer must settle for actually receiving a multitude of skewed *physical* clocks. To achieve low clock skew, it is important to carefully match all of the paths in the clock network and to minimize the delay through the network because random variations lead to skews that are a fraction of the mismatched delays. Previously, we have focused on hiding skew where possible and budgeting where necessary. We must be careful, however, that our skew-tolerant circuit techniques do not complicate the clock network so much that they introduce more skew than they tolerate.

This chapter begins by defining the nominal waveforms of physical clocks. The interarrival time of two clock edges is the delay between the edges. Clock skew is the absolute difference between the nominal and actual interarrival times of a pair of physical clock edges. Clock skew displays both spatial and temporal locality; by considering such locality, we must budget or hide only the actual skew experienced between launching and receiving clocks of a particular path. Skew budgets for min-delay checks must be more conservative than for max-delay because of the dire consequences of hold time violations; fortunately, min-delay races are set by pairs of clocks sharing a common edge in time, so min-delay budgets need not include jitter or duty cycle variation. Because it may be impractical to tabulate the skew between every pair of physical clocks on a chip, we lump clocks into *domains* for simplified, though conservative, analysis.

Having defined clock skew, we turn to skew-tolerant domino clock generation schemes for two, four, and more phases. We see that the clock generators introduce additional delays into the clock network and hence increase clock skew. Nevertheless, the extra clock skew is small compared to the skew tolerance, so such generators are acceptable. Four-phase skew-tolerant domino proves to be a reasonable design point combining good skew tolerance and simple clock generation, so we present a complete four-phase clock generation network supporting clock enabling and scan.

5.1   Clock Waveforms

We have relied upon an intuitive definition of clock skew while discussing skew-tolerant circuit techniques. In this chapter, we will develop a more precise definition of clock skew that takes advantage of the myriad correlations between physical clocks. Physical clocks may have certain systematic timing offsets caused by different numbers of clock buffers, clock gating, and so on. We can plan for these systematic offsets by placing more logic in some phases and less in others than we would have if all physical clocks exactly matched the logical clocks; the nominal offsets between physical clocks do not appear in our skew budget. The only term that must be budgeted as skew is the variability, the difference between nominal and actual interarrival times of physical clocks.

5.1.1   Physical Clock Definitions

A system has a small number of logical clocks. For example, flip-flops or pulsed latches use a single logical clock, transparent latches use two logical clocks, and skew-tolerant domino uses N, often four, logical clocks. A logical clock arrives at all parts of the chip at exactly the same time. Of course logical clocks do not exist, but they are a useful fiction to simplify design.

Conceptually, we envision a unique physical clock for each latch, but we can quickly group physical clocks that represent the same logical clock and have very small skew relative to each other into one clock to reduce the number of physical clocks. For example, a single physical clock might serve a bank of 64 latches in a datapath. By defining waveforms for physical clocks rather than logical clocks, we set ourselves up to budget only the skew actually possible between a pair of physical clocks rather than the worst-case skew experienced across the chip.

We define the set of physical clocks to be $C = \{\phi_1, \phi_2, \ldots, \phi_k\}$. We assume that all clocks have the same cycle time T_c.[1] Variables describing

---

1. In extremely fast systems, clocks may operate at very high frequency in local areas, but at lower frequency when communicating between remote units. We presently see this in systems where the CPU operates at high speed but the motherboard operates at a fraction of the frequency. This clocking analysis may be generalized to clocks with different cycle times.

the clock waveforms are defined below and illustrated in Figure 5.1 for a two-phase system with four 50% duty cycle physical clocks.

- T_c: the clock cycle time, or period

- T_{ϕ_i}: the duration for which ϕ_i is high

- s_{ϕ_i}: the start time, relative to the beginning of the common clock cycle, of ϕ_i being high

- $S_{\phi_i\phi_j}$: a phase shift operator describing the difference in start time from ϕ_i to the next occurrence of ϕ_j. $S_{\phi_i\phi_j} \equiv s_{\phi_i} - (s_{\phi_j} + WT_c)$, where W is a wraparound variable indicating the number of cycle crossings between the sending and receiving clocks. W is 0 or 1 except in systems with multicycle paths. Note that $S_{\phi_i\phi_i} = -T_c$ because it is the shift between consecutive rising edges of clock phase ϕ_i.

Note that Figure 5.1 labels the clocks $C = \{\phi_{1a}, \phi_{1b}, \phi_{2a}, \phi_{2b}\}$ rather than $C = \{\phi_1, \phi_2, \phi_3, \phi_4\}$ to emphasize that the four physical clocks correspond to only two logical clocks. The former labeling will be used for examples, while the latter notation is more convenient for expressing constraints in timing analysis in Chapter 6. The phase shifts between these clocks seen at consecutive transparent latches are shown in Table 5.1. Notice that the systematic offsets between clocks appear as different phase shifts rather than clock skew. It is possible to design around such systematic offsets, intentionally placing more logic in one half-cycle than another. Indeed,

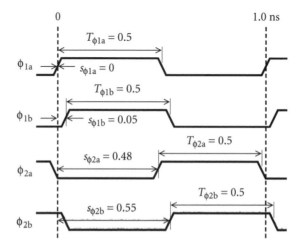

Figure 5.1   Two-phase clock waveforms

Table 5.1 Phase shift between clocks of Figure 5.1

		Receiving clock ϕ_i			
S_{ϕ_i,ϕ_j}		ϕ_{1a}	ϕ_{1b}	ϕ_{2a}	ϕ_{2b}
Launching clock ϕ_i	ϕ_{1a}			-0.48	-0.55
	ϕ_{1b}			-0.43	-0.50
	ϕ_{2a}	-0.52	-0.57		
	ϕ_{2b}	-0.45	-0.50		

designers sometimes intentionally delay clocks to extend critical cycles of logic in flip-flop–based systems where time borrowing is not possible. We save the term "skew" for uncertainties in the clock arrival times.

5.1.2 Clock Skew

If the actual delay between two phases ϕ_i and ϕ_j equalled the nominal delay S_{ϕ_i,ϕ_j}, the phases would have zero skew. Of course, delays are seldom nominal, so we must define clock skew. There are many sources of clock skew. When a single physical clock serves multiple clocked elements, delay between the clock arrivals at the various elements appears as skew. Cross-die variations in processing, temperature, and voltage also lead to skew. Electromagnetic coupling and load capacitance variations [18] lead to further skew in a data-dependent fashion. If all clock paths sped up or slowed down uniformly, the interarrival times would be unaffected and no skew would be introduced. Therefore, we are only concerned with differences between delays in the clock network.

In previous chapters, we have used a single skew budget t_{skew} that is the worst-case skew across the chip, in other words, the largest absolute value of the difference between the nominal and actual interarrival times of a pair of clocks anywhere on the chip. When t_{skew} can be held to about 10% of the cycle time, it is simple and not overly limiting to budget this worst-case skew everywhere. As skews are increasing relative to cycle time, we would prefer to only budget the actual skew encountered on a path, so we define skews between specific pairs of physical clocks. For example, $t_{skew}^{\phi_i,\phi_j}$ is the skew between ϕ_i and ϕ_j, the absolute value of the difference between the nominal and actual interarrival times of these edges measured at any pair of elements receiving these clocks. For a given pair of clocks, certain transitions may have different skews than others. Therefore, we also define skews between particular edges of pairs of physical clocks. For example, $t_{\text{skew}}^{\phi_i(r),\,\phi_j(f)}$ is the skew between the rising edge of ϕ_i and the falling edge of ϕ_j. $t_{\text{skew}}^{\phi_i,\,\phi_j}$ is the maximum of the skews between any edges of the clocks.

Notice that skew is a positive difference between the actual and nominal interarrival times, rather than being plus or minus from a reference point. When using this information in a design, we assume the worst: for maximum-delay (setup time) checks, that the receiving clock is skewed early relative to the launching clock; and for minimum-delay (hold time) checks, that the receiving clock is skewed late relative to the launching clock. If skews are asymmetric around the reference point, we may define separate values of skew for min- and max-delay analysis.

Also, note that the cycle count between edges is important in defining skew. For example, the skew between the rising edge of a clock and the same rising edge a cycle later is called the *cycle-to-cycle jitter*. The skew between the rising edge and the same rising edge many cycles later may be larger and is called the *peak jitter*. Generally, we will only consider edges separated by at most one cycle when defining clock skew because including peak jitter is overly pessimistic. This occasionally leads to slightly optimistic results in latch-based paths in which a signal is launched on the rising edge of one latch clock and passes through more than one cycle of transparent latches before being sampled. The jitter between the launching and sampling clocks is greater than cycle-to-cycle jitter in such a case, but the error is unlikely to be significant.

Since clock skew depends on mismatches between nominally equal delays through the clock network, skews budgets tend to be proportional to the absolute delay through the network. Skews between clocks that share a common portion of the clock network are smaller than skews between widely separated clocks because the former clocks experience no environmental or processing mismatches through the common section. However, even two latches sharing a single physical clock experience cycle-to-cycle skew from jitter and duty cycle variation, which depend on the total delay through the clock network.

The designer may use different skew budgets for minimum- and maximum-delay analysis purposes. Circuits with hold time problems will not operate correctly at any clock frequency, so designers must be very conservative. Fortunately, min-delay races occur between clocks in a single cycle, so jitter and duty cycle variation are not part of the skew budget. Circuits with setup time problems operate properly at reduced frequency. Therefore, the designer may budget an expected skew, rather than a worst-case skew, for max-delay analysis, just as designers may target TT processing instead of SS processing. This avoids overdesign while achieving acceptable yield at the target frequency. Unfortunately, calculating the expected skew requires extensive statistical knowledge of the components of clock skew and their correlations.

On account of larger chips, greater clock loads, and wire delays that are not scaling as well as gate delays, it is very difficult to hold clock skew across the die constant in terms of gate delays. Indeed, Horowitz predicted that keeping global skews under 200 ps is hard [39]. Moreover, as cycle times measured in gate delays continue to shrink, even if clock skew were held constant in gate delays, it would tend to become a larger fraction of the cycle time. Therefore, it will be very important to take advantage of skew-tolerant circuit techniques and to exploit locality of clock skew when building fast systems.

5.1.3 Clock Domains

Although you may conceptually specify an array of clock skews between each pair of physical clocks in a large system, such a table may be huge and mostly redundant. In practice, designers usually lump clocks into a hierarchy of clock domains. For example, we have intuitively discussed local and global clock domains; pairs of clocks in a particular local domain experience local skew, which is smaller than the global skew seen by clocks in different domains. We can extend this notion to finer granularity by defining a skew hierarchy with more levels of clock domains, as shown in Figure 5.2 for a system based on an H-tree.

In Figure 5.2, level 1 clock domains contain a single physical clock. Therefore, two elements in the level 1 domain will only see skew from RC delays along the clock wire and from jitter of the physical clock. Level 2 clock domains contain a clock and its complement and see additional skew caused by differences between the nominal and actual clock generator delays. Remember that systematic delay differences that are predictable at design time can be captured in the physical clock waveforms; only delay differences from process variations or environmental differences between the clock generators appear as skew. Higher-level clock domains

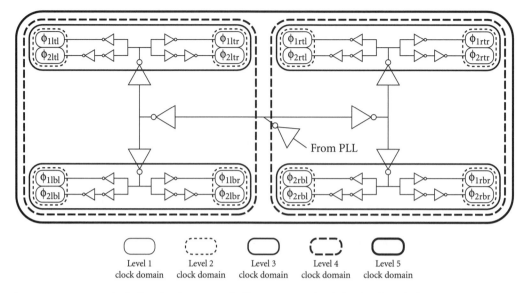

Figure 5.2   H-tree clock distribution network illustrating multiple levels of clock domains

see progressively more skew as delay variations in the global clock distribution network appear as skew.

5.2   Skew-Tolerant Domino Clock Generation

In most high-frequency systems, a single clock gclk is distributed globally using a modified H-tree or grid to minimize skew. Skew-tolerant domino can use this same clock distribution scheme with a single global clock. Within each unit or functional block, local clock generators produce the multiple phases required for latches and skew-tolerant domino. These local generators inevitably increase the delay of the distribution network and hence increase clock skew. This section describes several local clock generation schemes and analyzes the skews introduced by the generators. The simplest schemes involve simple delay lines and are adequate for many applications. Lower skews can be achieved using feedback systems with delays that track with process and environmental changes. We conclude with a full-featured local clock generator supporting transparent latches and four-phase skew-tolerant domino, clock enabling, and scan.

5.2.1 Delay Line Clock Generators

Overlapping clocks for skew-tolerant domino can be easily generated by delaying one or both edges of the global clock with chains of inverters. Figure 5.3 shows a simple two-phase skew-tolerant domino local generator, while Figure 5.4 extends the design to support four phases. The two-phase design uses a low-skew complement generator to produce complementary signals from the global clock. For example, Shoji showed how to match the delay of two and three inverters independently of relative PMOS/NMOS mobilities [78]. The falling clock edges are stretched with clock choppers to produce waveforms with greater than 50% duty cycle. Using a fanout of 3–4 on each inverter delay element sets reasonable delay

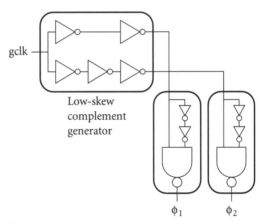

Figure 5.3 Two-phase clock generator

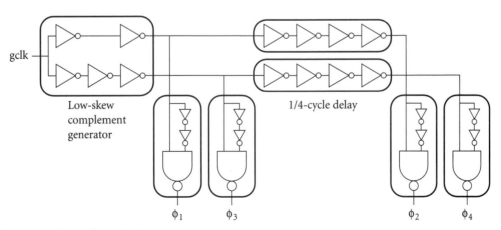

Figure 5.4 Four-phase clock generator

and minimizes the area of the clock buffer while preserving sharp clock edges.

The four-phase design is very similar, but uses an additional chain of inverters to produce a nominal quarter-cycle delay. At first it would seem such a clock generator would suffer from terrible clock skews because between best- and worst-case processing and environment, its delay may vary by a factor of two! Fortunately, we are concerned not with the absolute delay of the inverter chain, but rather with its tracking relative to critical paths on the chip. In the slow corner, the delay chain will have a greater delay, but the critical paths will also be slower and the operating frequency will be correspondingly reduced. Hence, to first order, the delay chain tracks the speed of the logic on the chip; we are now concerned about skew introduced by second-order mismatches.

Local-Generator–Induced Clock Skew

Since the local generators are not replicas of the circuits they are tracking, and indeed are static gates tracking the speed of dynamic paths, their relative delays may vary over process corners as well as due to local variation in voltage, temperature, and processing. Simulation finds that when most of the chip is operating under nominal processing and worst-case environment but a local clock generator sees a temperature 30°C lower and supply voltage 300 mV higher, the local generator will run 13% faster than nominal (6% from temperature, 7% from voltage). The relative delay of simple domino gates with respect to FO4 inverters varies up to about 6% across process corners. Finally, process tilt (i.e., fluctuation in L_e, t_{ox}, etc., across the die) may speed the local clock generator more than nearby logic. Little data is available on process tilt, but if we guess it causes a similar 13% variation, we conclude that nearly a third of the total local clock generator delay appears as clock skew.

Four-phase clock generators have a quarter-cycle more delay than two-phase generators, so are subject to more skew. However, they can also tolerate a quarter-cycle more skew than their two-phase counterparts, which is significantly more than the extra skew of the generators. For example, consider two- and four-phase systems like those described in Section 3.1.2 with cycle times of 16 FO4 delays and precharge times of 4 FO4 delays. If the local skew is 1 FO4 delay, the nominal overlap between phases is 3 FO4 delays for the two-phase system and 7 FO4 delays for the four-phase system. These overlaps can be used to tolerate clock skew and allow time borrowing. From the overlap we must subtract the skews introduced by the local clock generators. If the complement generator, clock chopper, and quarter-cycle delay lines have nominal delays of 2, 3, and 4 FO4 delays, respectively, we must budget 32% of these delays as additional skew. Figure 5.5 compares the remaining overlaps of each system, showing that although the four-phase system pays a larger skew penalty, the remaining overlap is proportionally much greater than that of the two-phase system. The four-phase clock generator can be simplified to use 50% duty cycle clocks as shown in Figure 5.6, eliminating the clock choppers at the expense of smaller overlaps. The four-phase system with 50% duty cycle waveforms still provides more overlap than the two-phase system and avoids the min-delay problems associated with overlapping two-phase clocks. Therefore, it is a reasonable design choice, especially

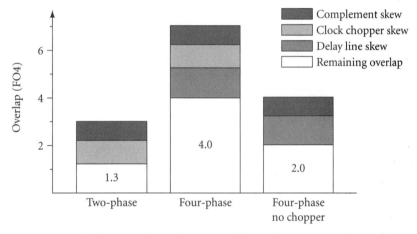

Figure 5.5 Overlap between phases for two- and four-phase systems after clock generator skews

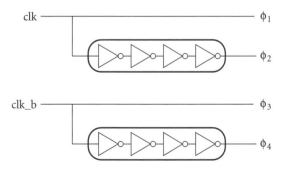

Figure 5.6 Simplified four-phase clock generator

considering the drawbacks of clock choppers that we will shortly note. In Section 5.2.3 we will look at a complete four-phase clock generator including clock gating and scan capability.

The four-phase clock generator with clock choppers appears to offer substantial benefits over the design with no choppers. A closer look reveals several liabilities in the design with choppers. Variations in the clock chopper delay cause duty cycle errors that cut into the precharge time, necessitating smaller overlaps than our first-order analysis predicted. The extended duty cycle also increases the susceptibility to min-delay problems, especially when coupled with the large skews introduced by the clock generator. Finally, the designer may still desire to use 50% duty cycle clocks for transparent latches. Therefore, the chopperless four-phase scheme is preferred when it offers enough overlap to handle the expected skews and time-borrowing requirements.

In addition to having adequate overlap for time borrowing and hiding clock skew, domino clocks must have sufficiently long precharge times in all process corners. The local clock generators are subject to duty cycle variation, which might change the amount of time available for precharging. Fortunately, if we design the system to have adequate precharge time in the worst-case environment under TT processing, environmental changes will only lead to more precharge time and faster precharge operation. In the SS corner, the clock must be slowed to accommodate precharge, but it is slowed anyway because of the longer critical paths.

N-Phase Local Clock Generators

Another popular skew-tolerant domino clocking scheme is to provide one phase for each gate. This offers maximum skew tolerance and more precharge time, as discussed in Section 3.1.4, at the expense of generating and distributing more clocks and roughly matching gate and clock delays. Figures 5.7 and 5.8 show such clock generation schemes. Figure 5.7 uses both edges of the clock and is the simplest scheme. The exact delay of the buffers is not particularly important so long as the clocks arrive at each gate before the data. Figure 5.8 delays a single clock edge, as used on the IBM guTS experimental GHz processor [62, 86], its successor [38], and on the Sun UltraSparc III [32]. To make sure the last phase overlaps the first phase of the next cycle, a pulse stretcher, such as an SR latch, must be used. The stretcher is especially important at low frequency; the first guTS test chip accidentally omitted a stretcher, making the chip only run at a

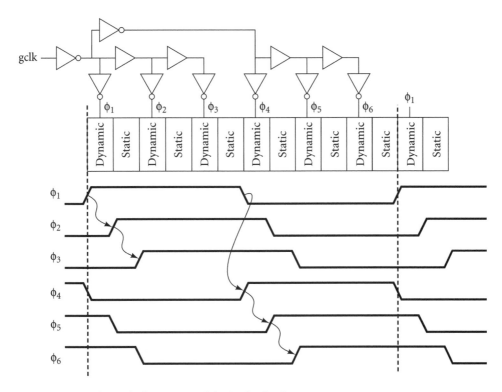

Figure 5.7 *N*-phase clock generator delaying both edges

narrow range of frequencies. Another disadvantage of delaying a single edge is that the precharge time of the last phase becomes independent of clock frequency, creating another timing constraint that cannot be fixed by slowing the clock. Finally, the longer delays of the single-edge design lead to greater clock skew. Therefore, the design delaying both edges is simpler and more robust.

5.2.2 Feedback Clock Generators

To reduce the skew and duty cycle uncertainty of the local clock generators, we may also use local delay-locked loops [14] to produce skew-tolerant domino clocks. Such a system is shown in Figure 5.9. The local loop receives the global clock and delays it by exactly one quarter-cycle by adjusting a delay line to have a half-cycle delay and tapping off the middle of the delay line. The feedback controller compensates for process and low-frequency environmental variations and even for a modest range of different clock frequencies. The art of DLL design is beyond

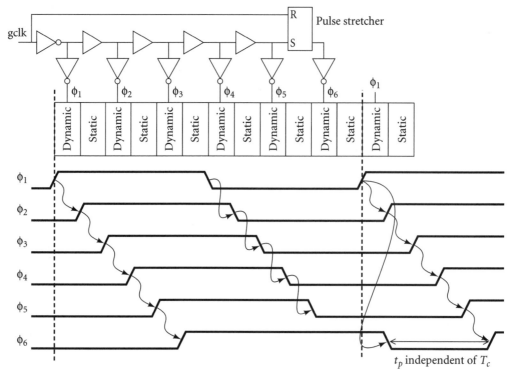

Figure 5.8 *N*-phase clock generator delaying a single edge

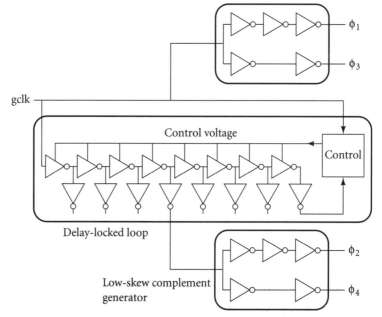

Figure 5.9 Four-phase clock generator using feedback control

the scope of this work; the illustration should be considered conceptual only.

Unfortunately, the DLL itself introduces skew. In particular, power supply noise in the delay line at frequencies above the controller bandwidth appears as jitter on ϕ_2 and ϕ_4. In a system without feedback, power supply variation from $V + \Delta V$ to $V - \Delta V$ causes delay variation from $t + \Delta t$ to $t - \Delta t$. In the DLL, a supply step from $V + \Delta V$ to $V - \Delta V$ after the system had initially stabilized at $V + \Delta V$ causes delay variation from t to $t - 2\Delta t$. Similarly, a rising step causes delay variation from t to $t + 2\Delta t$. Therefore, the DLL has twice the voltage sensitivity of the system without feedback. PLLs are even more sensitive to voltage noise because they accumulate jitter over multiple cycles; therefore, they are not a good choice for local clock generators.

Fortunately, the local high-frequency voltage noise causing jitter is a fraction of the total voltage noise. If we assume the high-frequency noise in each DLL is half as large as the total voltage noise, the jitter of the DLL will equal the skew introduced by voltage errors on a regular delay line system. Using the numbers from the example in Section 5.2.1, this corresponds to 7% of the quarter-cycle delay to the line tap. The local clock generator also is subject to variations in the complement generator. If the DLL is designed to achieve negligible static phase offset, the skew improvement of the feedback system over the delay line system is predicted to be the difference in delay sensitivity, 32% − 7%, times the quarter cycle delay, or about 6% of the cycle time. This comes at the expense of building a small DLL in every local clock generator. The DLL may use an improved delay element with reduced supply sensitivity, but the same delay elements may be used in ordinary delay lines. The designer must weigh the potential skew improvement of DLL-based clock generators against the area, power, and design complexity they introduce. In today's systems, simple delay lines are probably good enough, but in future systems with even tighter timing margins, DLLs may offer enough advantages to justify their costs.

5.2.3   Putting It All Together

So far we have only considered generating periodic clock waveforms. Most systems also require the ability to stop the clock and to scan data in and out of each cycle. We saw in Section 4.3.2 that scan required precise release of the scan enable signal. By building the release circuitry into the clock generator, we avoid the need to route timing-critical global scan signals. In this section we integrate such scan circuitry and clock enabling with four-phase skew-tolerant domino to illustrate a complete local clock generator.

Local clocks are often gated with an enable signal to produce a qualified clock. Qualified clocks can be used to save power by disabling inactive units, to prevent latching new data during a pipeline stall, or to build combined multiplexer-latches. Clock enable signals are often critical because they come from far away or are data dependent. Therefore, it is useful to minimize the setup time of the clock enable before the clock edge.

Figure 5.10 illustrates a complete local clock generator for a four-phase skew-tolerant domino system. It receives gclk from global clock distribution network and an enable signal for the local logic block. It generates the four regular clock phases along with a variant of ϕ_4 used for scan. Different clock enables can be used for different gates or banks of gates as appropriate. Using a two-input NAND gate in all local clock generators provides best matching between phases to minimize clock skew; the enable may be tied high on some clocks that never stop. The last domino gate in each cycle uses ϕ_4 for precharge and ϕ_{4s} for evaluation. Two-phase static latches use ϕ_1 and ϕ_3 as clk and clk_b. The clock generator uses delay chains to produce domino phases ϕ_2 and ϕ_4 delayed by one quarter of the nominal clock period. Scan is built into static latches and domino gates as described in Section 4.3. Notice that when *SCA* is asserted, an SR latch forces ϕ_{4s} low to disable the dynamic gate being scanned. When ϕ_4 falls to begin precharge, the SR latch releases ϕ_{4s} to resume normal operation. Therefore, we avoid distributing any high-speed global scan enable signals and can use exactly the same scan procedure as we used with static latches:

1 Stop gclk low.

2 Toggle *SCA* and *SCB* to march data through the scan chain. The first pulse of *SCA* will force ϕ_{4s} low.

3 Restart gclk. The falling edge of ϕ_4 will release ϕ_{4s} to track ϕ_4.

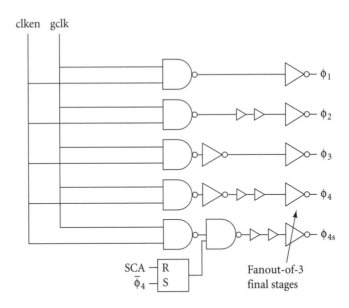

Figure 5.10 Local four-phase clock generators supporting scan and clock enabling

5.3 Summary

Circuit designers wish to receive a small number of logical clocks simultaneously at all points of the die. They must instead accept a huge number of physical clocks arriving at slightly different times to different receivers. Clock skew is the difference between the nominal and actual interarrival times of two clocks. It depends on numerous sources that are difficult or impossible to model accurately, so it is typically budgeted using conservative engineering estimates. Because clock skew is an increasing problem, it is important to understand the sources and avoid unnecessary conservatism. Skew budgets therefore may depend on the phases of and distance between the physical clocks, the particular edges of the clocks, and the number of cycles between the edges. Clocks may be grouped into a hierarchy of clock domains according to their relative skews; communication within a domain experiences less skew than communication across domains.

The designer has three tactics to deal with skew: budget, hide, and minimize. Taking advantage of the smaller amounts of skew between nearby elements is a powerful way to minimize skew, but requires improved timing analysis algorithms, which are the subject of Chapter 6.

Skew-tolerant circuit techniques hide clock skew, but the local clock generators necessary to produce multiple overlapping clock phases for skew-tolerant domino introduce skew of their own. Fortunately, the skews introduced are less than the tolerance provided, so skew-tolerant domino is an overall improvement.

5.4   Exercises

[15]   **5.1**   What is the distinction between physical and logical clocks? How many logical clocks exist in Figure 5.2? How many physical clocks?

[15]   **5.2**   Why is it useful to distinguish between systematic and random variations in the start time of two physical clocks corresponding to the same logical clock? How can the designer use this information to avoid being overly pessimistic in design?

[15]   **5.3**   What are some advantages of separately defining skews between pairs of clocks rather than providing a single global skew number? What are some disadvantages?

[15]   **5.4**   Why is the pulse stretcher in Figure 5.8 required? Draw a timing diagram to explain how the circuit might fail if the pulse stretcher were omitted.

[15]   **5.5**   What is the function of the SR latch in Figure 5.10? Why is it preferable to use the SR latch rather than providing a special scan enable signal to the second NAND gate in the ϕ_{4s} generator?

[15]   **5.6**   Many CAD papers describe algorithms for generating "zero-skew" clock trees (e.g., [88]). What is misleading about the term "zero-skew" from the designer's point of view? What term would you use instead?

6
Timing Analysis

It is impractical to build complex digital systems without CAD tools to analyze and verify the designs. Therefore, novel circuit techniques are of little use without corresponding CAD tools. Although most standard circuit tools such as simulators, layout-versus-schematic checkers, ERC and signal integrity verifiers, and so on, work equally well for skew-tolerant and non–skew-tolerant circuits, timing analyzers must be enhanced to understand and take advantage of different amounts of clock skew between different clocks.

Timing analysis addresses the question of whether a particular circuit will meet a timing specification. The analysis must check maximum delays to verify that a circuit will meet setup times at the desired frequency, and minimum delays to verify that hold times are satisfied. This chapter describes how to extend a traditional formulation of timing analysis to handle clock skew, including different budgets for skew between different regions of a system.

Our formulation of timing analysis is built on an elegant framework from Sakallah, Mudge, and Olukotun [70] for systems with transparent latches. Although the framework assumes zero clock skew, we can easily support systems with a single clock domain by adding worst-case skew to the setup time of each latch. We then develop an exact set of constraints for analyzing systems with different amounts of skew between different elements. This exact analysis leads to an explosion of the number of timing constraints. By introducing a hierarchy of clock domains with tighter bounds on skews within smaller domains, we offer an approximate analysis that is conservative, but less pessimistic than the single skew scenario and with many fewer constraints than the exact analysis. Once we understand how to analyze latches in a system with multiple clock domains, we find analyzing flip-flops is even easier. Domino gates also fit nicely into the framework, sometimes behaving as latches and sometimes as flip-flops. Having solved the problem of max-delay, we show that min-delay is much easier to check because it does not involve time borrowing. We conclude by presenting algorithms for verifying the timing constraints and showing that, for a large test case, the exact analysis is only slightly more expensive than the skewless analysis.

6.1 Timing Analysis without Clock Skew

We will begin by describing a formulation of timing analysis for latch-based systems from Sakallah et al. [70]. The simplicity of the formulation stems from a careful choice of time variables describing data inputs and outputs of the latches. In this section, we consider only D-type latches with data in, data out, and clock terminals. Section 6.3 extends the model to include other clocked elements such as flip-flops and domino gates.

A system contains a set of physical clocks $C = \{\phi_1, \phi_2, \ldots, \phi_k\}$ with a common cycle time T_c, and a set of latches $L = \{L_1, L_2, \ldots, L_l\}$. As defined in Section 5.1.1, the clocks have a duration T_{ϕ_i}, start time s_{ϕ_i}, and phase shift operator $S_{\phi_i\phi_j}$. For each of the l latches in the system, we define the following variables and parameters that describe which clock is used to control each latch, when data arrives and departs each latch, and the setup time and propagation delay of each latch:

- p_i: the clock phase used to control latch i

- A_i: the arrival time, relative to the start time of p_i, of a valid data signal at the input to latch i

- D_i: the departure time, relative to the start time of p_i, at which the signal available at the data input of latch i starts to propagate through the latch

- Q_i: the output time, relative to the start time of p_i, at which the signal at the data output of latch i starts to propagate through the succeeding stages of combinational logic

- Δ_{DCi}: the setup time for latch i required between the data input and the trailing edge of the clock input

- Δ_{DQi}: the maximum propagation delay of latch i from the data input to the data output while the clock input is high

Finally, define the propagation delays between pairs of latches:

- Δ_{ij}: the maximum propagation delay through combinational logic between latch i and latch j. If there are no combinational paths from latch i to latch j, $\Delta_{ij} \equiv -\infty$ effectively eliminates the path from consideration.

Using these definitions we can express constraints on the propagation of signals between latches and the setup of signals before the sampling edges of the latches. Setup time constraints require that a signal arrive at a latch some setup time before the sampling clock edge. Thus:

$$\forall i \in L \qquad A_i + \Delta_{DCi} \le T_{p_i} \tag{6.1}$$

The propagation constraints relate the departure, output, and arrival times of latches. Data departs a latch input when the data arrives and the latch is transparent:

$$\forall i \in L \qquad D_i = \max(0, A_i) \tag{6.2}$$

The latch output becomes valid some latch propagation delay after data departs the input.

$$\forall i \in L \qquad Q_i = D_i + \Delta_{DQi} \tag{6.3}$$

Finally, the arrival time at a latch is the latest of the possible arrival times from data leaving other latches and propagating through combinational logic to the latch of interest. Notice that the phase shift operator S must be added to translate between relative times of the launching and receiving latch clocks.

$$\forall i, j \in L \qquad A_i = \max(Q_j + \Delta_{ji} + S_{p_j p_i}) \tag{6.4}$$

Observe that both D_i and Q_i will always be nonnegative quantities because a signal may not begin propagating through a latch until the clock has risen. A_i is unrestricted in sign because the input data may arrive before or after the latch clock. By assuming that clock pulse widths T_i are always greater than latch setup times Δ_{DCi} and eliminating the Q and A variables, we can rewrite these constraints as L1 and L2 exclusively in terms of signal departure times and the clock parameters.

L1. Setup Constraints:

$$\forall i \in L \qquad D_i + \Delta_{DCi} \le T_{p_i} \tag{6.5}$$

L2. Propagation Constraints:

$$\forall i, j \in L \qquad D_i = \max(0, \max \langle D_j + \Delta_{DQj} + \Delta_{ji} + S_{p_j p_i} \rangle) \tag{6.6}$$

From these constraints, one can either verify that a design will operate at a target frequency or compute the maximum possible frequency at which the design functions. Szymanski and Shenoy present a relaxation algorithm for the timing verification problem [85], while Sakallah et al. reformulate the constraints as a linear program for cycle time optimization [70]. We will return to these algorithms in Section 6.5. In the meantime, let us consider an example to get accustomed to the notation.

EXAMPLE 6.1   Consider the microprocessor core shown in Figure 6.1. The circuit consists of two clocks $\{\phi_1, \phi_2\}$ and five latches $\{L_3, L_4, \ldots, L_7\}$ with logic blocks with propagation delays $\Delta 4$ through $\Delta 7$. Latches L_4 and L_5 comprise the ALU, while L_6 and L_7 comprise the data cache. The ALU results may be bypassed back for use on a subsequent ALU operation or may be sent as an input to the data cache. The data cache output may be returned as an input to the ALU. Assume that the latch setup time Δ_{DCi} and propagation delay Δ_{DQi} are 0, and that the external input to L_3 arrives early enough so that it can depart the latch at $D_3 = 0$. The clocks have a target cycle time $T_c = 10$ units and 50% duty cycle, giving phase length $T_p = T_{\phi 1} = T_{\phi 2} = T_c/2$. Write all of the setup and propagation constraints. What are the latch departure times if the logic delays are $\Delta 4 = 5; \Delta 5 = 5; \Delta 6 = 5; \Delta 7 = 5$? If the logic delays are $\Delta 4 = 7; \Delta 5 = 3; \Delta 6 = 5; \Delta 7 = 4$?

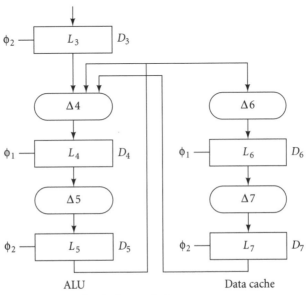

Figure 6.1   Example circuit for timing analysis

SOLUTION The complete set of timing constraints are listed as the skewless formulation in Appendix A.

In the first case, the departure times are $D_4 = 0$; $D_5 = 0$; $D_6 = 0$; $D_7 = 0$. This case illustrates perfectly balanced logic. Each combinational block contains exactly half a cycle of logic. Data arrives at each latch just as it becomes transparent. Therefore, all the departure times are 0.

In the second case, the departure times are $D_4 = 2$; $D_5 = 0$; $D_6 = 0$; $D_7 = 0$. This case illustrates time borrowing between half-cycles. Combinational block 4 contains more than half a cycle of logic, but it can borrow time from combinational block 5 to complete the entire ALU operation in one cycle. Combinational block 7 finishes early, but cannot depart latch 7 until the latch becomes transparent; this is known as clock blocking. The positive departure time indicates the time borrowing through L_4. ∎

6.2 Timing Analysis with Clock Skew

Recall from Section 5.1.1 that a system has a small number of logical clocks, but possibly a much greater number of skewed physical clocks. Sakallah's formulation, discussed in the previous section, does not account for clock skew; in other words, it assumes that all clocked elements receive their ideal logical clock. Because clock skews are becoming increasingly important, we now examine how to include skew in timing analysis. We first describe a simple modification to the setup constraints that accounts for a single clock skew budget across the chip. Unfortunately, this is very pessimistic because most clocked elements see much less than worst-case skew. Next we develop an exact analysis allowing for different skews between each pair of clocks. This leads to an explosion in the number of timing constraints for large circuits. By making a simple approximation of clock domains, we finally formulate the problem with fewer constraints in a way that is conservative, yet less pessimistic than the single-skew approach.

To illustrate systems with clock skew, we use a more elaborate model of the ALU from Figure 6.1. Our new model, shown in Figure 6.2, contains clocks $C = \{\phi_{1a}, \phi_{1b}, \phi_{2a}, \phi_{2b}\}$, where physical clocks ϕ_{1a} and ϕ_{1b} are nominally identical to logical clock ϕ_1, but are located in different parts of

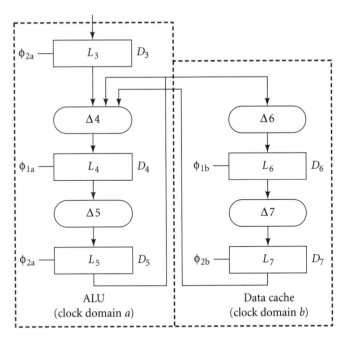

Figure 6.2   Example circuit with clock domains

the chip and subject to skew. Only a small $t_{\text{skew}}^{\text{local}}$ exists between clocks in the same domain, but the larger $t_{\text{skew}}^{\text{global}}$ may occur between clocks in different domains.

6.2.1   Single Skew Formulation

The simplest and most conservative way to accommodate clock skew in timing analysis is to use a single upper bound on clock skew. Suppose that we assume a worst-case amount of clock skew, $t_{\text{skew}}^{\text{global}}$, may exist between any two clocked elements on an integrated circuit. Shenoy [73] shows that such skew can be accommodated in the analysis by modifying the setup time constraint. Data must set up before the falling edge of the clock, yet there may be skew between launching and receiving elements such that the data was launched off a late clock edge and is sampled on an early edge. Therefore, we must add clock skew to the effective setup time:

L1S. Setup Constraints with Single Skew:

$$\forall i \in L \qquad D_i + \Delta_{DCi} + t_{\text{skew}}^{\text{global}} \leq T_{p_i} \qquad (6.7)$$

The propagation constraints are unchanged.

6.2.2 Exact Skew Formulation

In a real clock distribution system, clock skews between adjacent elements are typically much less than skews between widely separated elements. We can avoid budgeting global skew in all paths by considering the actual launching and receiving elements and only budgeting the possible skew that exists between the elements.

Unfortunately, the transparency of latches makes this a complex problem. Consider the setup time on a signal arriving at latch L_4 in Figure 6.2. How much skew must be budgeted in the setup time? The answer depends on the skew between the clock that originally launched the signal and the falling edge of ϕ_{1a}, the clock that is receiving the signal. For example, the signal might have been launched from L_7 on the rising edge of ϕ_{2b}, in which case $t_{\text{skew}}^{\phi_{2b}(r),\,\phi_{1a}(f)}$ must be budgeted. On the other hand, the signal might have been launched from L_5 on the rising edge of ϕ_{2a}, then propagated through L_6 and L_7 while both latches were transparent. In such a case, only the smaller skew $t_{\text{skew}}^{\phi_{2a}(r),\,\phi_{1a}(f)}$ must be budgeted because the launching and receiving clocks are in the same local domain despite the fact that the signal propagated through transparent elements in a different domain. We see that exact timing analysis with varying amounts of skew between elements must track not only the accumulated delay to each element, but also the clock of the launching element.

To track both accumulated delay and launching clock, we can define a vector of arrival and departure times at each latch, with one dimension per physical clock in the system. These times are still nominal, not including skew.

- A_i^c: the arrival time, relative to the beginning of p_i, of a valid data signal launched by clock c and now at the input to latch i

- D_i^c: the departure time, relative to the beginning of p_i, at which the signal launched by clock c and available at the data input of latch i starts to propagate through the latch

The setup constraints must budget the skew $t_{\text{skew}}^{c(r),\,p_i(f)}$ between the rising edge of the launching clock c and the falling edge of the clock p_i controlling the sampling element:

$$\forall i \in L,\, c \in C \qquad D_i^c + \Delta_{DCi} + t_{\text{skew}}^{c(r),\,p_i(f)} \leq T_{p_i} \tag{6.8}$$

The arrival time at latch i for a path launched by clock c depends on the propagation delay and departure times from other latches for signals also launched by clock c:

$$\forall i, j \in L, c \in C \qquad A_i^c = \max(D_j^c + \Delta_{DQj} + \Delta_{ji} + S_{p_jp_i}) \qquad (6.9)$$

If a latch is transparent when its input arrives, data should depart the latch at the same time it arrives and with respect to the same launching clock. If a latch is opaque when its input arrives, the path from the launching clock will never constrain timing and a new path should be started departing at time 0, launched by the latch's clock. Because of skew between the launching and receiving clocks, the receiving latch may be transparent even if the input arrives at a slightly negative time. To model this effect, we allow departure times with respect to a clock other than that which controls the latch to be negative, equal to the arrival times. Departure times with respect to the latch's own clock are strictly nonnegative. To achieve this, we define an identity operator I_{ϕ_1, ϕ_2} on a pair of clocks ϕ_1 and ϕ_2 that is the minimum departure time for a signal launched by one clock and received by the other: 0 if $\phi_1 = \phi_2$ and $-\infty$ if the clocks are different.

These setup and propagation constraints are summarized below. Notice that the number of constraints is proportional to the number of distinct clocks in the system. Also, notice that the constraints are orthogonal; there is no mixing of constraints from different launching clocks.

L1E. Setup Constraints with Exact Skew Analysis:

$$\forall i \in L, c \in C \qquad D_i^c + \Delta_{DCi} + t_{skew}^{c(r), p_i(f)} \leq T_{p_i} \qquad (6.10)$$

L2E. Propagation Constraints with Exact Skew Analysis:

$$\forall i, j \in L, c \in \qquad CD_i^c = \max(I_{c, p_i}, \max(D_j^c + \Delta_{DQj} + \Delta_{ji} + S_{p_jp_i})) \qquad (6.11)$$

A brief example may help explain negative departure times. Consider a path launched from L_6 in Figure 6.2 on the rising edge of ϕ_{1b}: $D_6^{\phi_{1b}} = 0$. Let the cycle time T_c be 10 units, and $t_{skew}^{\phi_{1b}, \phi_{2b}}$ be 1. Therefore, ϕ_{2b} may transition up to one unit of time earlier or later than nominal, relative to ϕ_{1b}, as shown in Figure 6.3. Also, suppose the latch propagation delay is 0, so $A_7^{\phi_{1b}} = \Delta 7 - 5$. If $\Delta 7$ is less than 4, the signal arrives at L_7 before the latch becomes transparent, even under worst-case clock skew. If $\Delta 7$ is

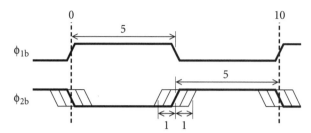

Figure 6.3 Clock waveforms including local skew

between 4 and 6 units, corresponding to $A_7^{\phi_{1b}}$ in the range of -1 to 1, the signal arrives at L_7 when the latch might be transparent, depending on the actual skew between ϕ_{1b} and ϕ_{2b}. If $\Delta 7$ is between 6 and 9 units, the signal arrives at L_7 when the latch is definitely transparent. Because the signal may depart the latch at the same time as it arrives when the latch is transparent, the departure time $D_7^{\phi_{1b}}$ may physically be as early as -1. We allow the departure time to be arbitrarily negative; if it is more negative than -1, it will always be less critical than the path departing L_7 on the rising edge of ϕ_{2b}. In Section 6.5, we will consider pruning departure times that are negative enough to always be noncritical for the sake of computational efficiency. Departure times must be nonnegative with respect to the clock controlling the latch; for example, $D_7^{\phi_{2b}} \geq 0$.

6.2.3 Clock Domain Formulation

The exact timing analysis formulation leads to an explosion in the number of constraints required for a system with many clocks; a system with k clocks has k times as many constraints as the single skew formulation. We would like to develop an approximate analysis that gives more accurate results than the single skew formulation, yet has fewer constraints than the exact formulation. To do this, we will formalize the concepts of skew hierarchies and clock domains.

A *skew hierarchy* is a collection of sets of clocks in the system. The sets are called *clock domains*. Each clock domain $d \subset C$ of the hierarchy has an associated number h called the *level* of the clock domain. A skew hierarchy has n levels, where level 1 clock domains are the smallest domains and the level n domain contains all the clocks C of the system. Define $H = \{1, \ldots, n\}$ to be the set of levels. To be a strict hierarchy, clock domains must not partially overlap; in other words, for any pair of clock domains, either one is a subset of the other or the domains are disjoint. If one domain contains another, the larger domain has the higher level. The case of $n = 1$ corresponds to assuming worst-case skew everywhere. The case of $n = 2$ is also interesting, corresponding to a system with local and global skews. We define the following skew hierarchy variables:

- t^h_{skew}: the upper bound on skew between two clocks in a level h clock domain. This quantity monotonically increases with h. The top level domain experiences global skew: $t^n_{skew} = t^{global}_{skew}$

- h_{ij}: the level of the smallest clock domain containing clocks i and j, that is, the minimum h such that $t^h_{skew} \geq t^{i,j}_{skew}$

We can also refer to skew between individual edges of clocks within a clock domain. For example, $t^{1(r,r)}_{skew}$ is the skew between rising edges of two clocks within a local clock domain. Because duty cycle variation occurs independently of clock domains, such skew between the same pair of edges is likely to be much smaller than the skew between different edges, such as $t^{1(r,f)}_{skew}$.

Skew hierarchies apply especially well to systems constructed in a hierarchical fashion. For example, Figure 5.2 illustrates an H-tree clock distribution network. It attempts to provide a logical two-phase clock consisting of ϕ_1 and ϕ_2 to the entire chip with zero skew. Although there are only two phases, the system actually contains 16 physical clocks for the purpose of modeling skew. All of the wire lengths in principle can be perfectly matched, so it is ideally possible to achieve zero systematic clock skew in the global distribution network. Even so, there is some RC delay along the final clock wires. Also, process and environmental variation in the delays of wires and buffers in the distribution network cause random clock skew. The clock skews between various phases depend on the level of their common node in the H-tree. For example, ϕ_{1ltl} and ϕ_{2ltl} only see a small amount of skew, caused by the final stage buffers and local routing. On the other hand, ϕ_{1ltl} and ϕ_{1rbr} on opposite corners of the chip may experience much more skew. The boxes show how the clocks could be collected into a five-level skew hierarchy.

The concept of skew hierarchies also applies to other distribution systems. For example, in a grid-based clock system, as used on the DEC Alpha 21164 [27], local skew is defined to be the RC skew between elements in a 500-micron radius, while global skew is defined to be the RC skew between any clocked elements on the die. Global skew is quoted at

90 ps, while local skew is only 25 ps.[1] Therefore, the chip could be partitioned into distinct 500-micron blocks so that elements communicating within blocks only see local skew, while elements communicating across blocks experience global skew.

The huge vector of timing constraints in the exact analysis is introduced because we track the launching clock of each path so that when the path crosses to another clock domain, and then returns to the original domain, only local skew must be budgeted at the latches in the original domain. An alternative is to only track whether a signal is still in the same domain as the launching clock or if it has ever crossed out of the local domain. In the first case, we budget only local clock skew. In the second case, we always budget global clock skew, even if the path returns to the original domain. This is conservative; for example, in Figure 6.2, a path that starts in the ALU, and then passes through the data cache while the cache latches are transparent and returns to the ALU, would unnecessarily budget global skew upon return to the ALU. However, it greatly reduces the number of constraints because we must only track whether the path should budget global or local skew, leading to only twice as many constraints as the single skew formulation. In general, we can extend this approach to handle n levels of hierarchical clock domains.

Again, we define multiple departure times, now referenced to the clock domain level of the signal rather than to the launching clock.

- A_i^h: the arrival time, relative to the beginning of p_i, of a valid data signal on a path that has crossed clock domains at level h of the clock domain hierarchy and is now at the input to latch i

- D_i^h: the departure time, relative to the beginning of p_i, at which the signal that has crossed clock domains at level h of the clock domain hierarchy and is now available at the data input of latch i starts to propagate through the latch

---

1. These skews sound very low, suggesting that clock skew is not a major problem for circuit designers. Unfortunately, published papers tend to quote only the systematic components of clock skew caused by RC delays and mismatched loads, not skew from process variations across the clock network or the time-varying components such as jitter from the generator and clock buffers. It is difficult to quantify and measure these time-varying components, but they do belong in a timing budget. *Caveat emptor.*

When a path crosses clock domains, it is bumped up to budget the greater skew; in other words, the skew level at the receiver is the maximum of the skew level of the launched signal and the actual skew level between the clocks of the departing and receiving latches. As usual, departure times with respect to the latch's own clock are strictly nonnegative, while departure times with respect to other clocks may be negative. Because we do not track the actual launching clock, but treat all clocks within a level 1 clock domain the same, we require that departure times from level 1 domains be nonnegative. To achieve this, we define an identity operator I_h on a level of the skew hierarchy that is the minimum departure time for a departure time at that level of the hierarchy: 0 for departures with respect to level 1, and $-\infty$ for departures with respect to higher levels.

The setup and propagation constraints are listed below. Notice that the number of constraints is now proportional only to the number of levels of the clock domain hierarchy, not the number of clocks or even the number of domains. For a system with two levels of clock domains (i.e., local and global) this requires only twice as many constraints as the single skew formulation.

L1D. Setup Constraints with Clock Domain Analysis:

$$\forall i \in L, h \in H \qquad D_i^h + \Delta_{DCi} + t_{\text{skew}}^{h(r,f)} \le T_{p_i} \tag{6.12}$$

L2D. Propagation Constraints with Clock Domain Analysis:

$$\forall i, j \in L, h_1 \in H, h_2 = \max(h_1, h_{p_i p_j})$$

$$D_i^{h_2} = \max(I_{h_2}, \max(D_j^{h_1} + \Delta_{DQj} + \Delta_{ji} + S_{p_j p_i})) \tag{6.13}$$

Yet another option is to lump clocks into a modest number of local clock domains, and then perform an exact analysis on paths that cross clock domains. The number of constraints in such an analysis is proportional to the number of local clock domains, which is smaller than the number of physical clocks required for exact analysis, but larger than the number of levels of clock domains. Paths within a local domain always budget local skew. This hybrid approach avoids unnecessarily budgeting global skew for paths that leave a local domain but return a receiver in the local domain.

6.2.4   Example

Let us return to the microprocessor example of Figure 6.2 to illustrate applying timing analysis to systems with four clocks and a two-level skew hierarchy. We will enumerate the timing constraints for each formulation, and then solve them to obtain minimum cycle time. This example will illustrate time borrowing, the impact of global and local skews, and the conservative approximations made by the inexact algorithms.

Suppose the nominal clock and latch parameters are identical to those in the example of Section 6.1, but that the system experiences $t_{\text{skew}}^{\text{local}} = 1$ of skew between clocks in a particular domain and $t_{\text{skew}}^{\text{global}} = 3$ of skew between clocks in different domains.

The timing constraints are tabulated in Appendix A for each formulation and were entered into a linear programming system. Various values of $\Delta 4$ to $\Delta 7$ were selected to test the analysis. The values were all selected so that a cycle time of 10 could be achieved in the case of no skew. The examples illustrate well-balanced logic, time borrowing between phases and across cycles, cycles limited by local and global skews, and a case in which the clock domain analysis yields conservative results.

Table 6.1 shows the values of combinational logic delay and cycle times achieved in each example. Bold data indicates conservative results caused by inexact analysis throwing away information. The clock domain results match the exact results in all cases but one, in which a path started in the ALU, passed through the cache while the latches were transparent, and returned to the ALU. Only local skew must be budgeted on return, but the clock domain analysis method conservatively budgeted global skew, leading to a pessimistic cycle time. The single skew formulation is conservative in three cases that used large amounts of time borrowing

Table 6.1   Examples of timing analysis results

$\Delta 4$	$\Delta 5$	$\Delta 6$	$\Delta 7$	T_c exact	T_c clock domains	T_c single skew	Notes
5	5	5	5	10	10	10	balanced logic
6	3	6	5	10	10	10	time borrowing
0.5	9.5	2.5	5	10.5	10.5	**12.5**	local skew limit
2	8	5	5	10.67	10.67	**11**	global skew limit
8	2	5	5	11	11	11	global skew limit
7	2	6	5	10	**10.5**	**10.5**	conservative result

where only local skew actually applied but global skew was unnecessarily budgeted.

6.3   Extension to Flip-Flops and Domino Circuits

So far, we have addressed the question of timing analysis for transparent latches. Pulsed latches have identical cycle time constraints as transparent latches and therefore are also handled. We can easily extend the framework to mix latches and edge-triggered flip-flops. Flip-flops are simpler because they do not allow time borrowing. We can also extend the framework to handle domino circuits, which may have the timing requirements of latches or flip-flops, depending on usage. The main change introduced in this section is to track both arrival and departure times because inputs to edge-triggered devices must arrive some setup time before the edge and do not depart until after the edge. We present only the exact analysis; the simplified formulation assuming clock domains is very similar.

6.3.1   Flip-Flops

For flip-flops, data must arrive before the rising edge of the clock phase, rather than the falling edge. Let $F = \{F_1, F_2, \ldots, F_f\}$ be the set of flip-flops. Data always departs the flop at the rising edge. We must therefore separately track arrival and departure times and introduce a set of departure constraints that relate arrival and departure times and nonnegativity. The setup and departure constraints are written differently for flip-flops and latches.

Setup Constraints for Flip-Flops:

$$\forall i \in F, c \in C \qquad A_i^c + \Delta_{DCi} + t_{\text{skew}}^{c(r), p_i(r)} \leq 0 \tag{6.14}$$

Note that the sampling edge for a flip-flop is the rising edge, so the skew is between two rising edges, rather than between the rising edge of the launching clock and the falling edge of the sampling clock as is the case for latches.

Setup Constraints for Latches:

$$\forall i \in L, c \in C \qquad A_i^c + \Delta_{DCi} + t_{\text{skew}}^{c(r),\, p_i(f)} \leq T_{p_i} \qquad (6.15)$$

Departure Constraints for Flip-Flops:

$$\forall i \in F \qquad D_i^{p_i} = 0 \qquad (6.16)$$

Note that there is no departure constraint from clocks other than the flop's launching clock because flip-flops are not transparent.

Departure Constraints for Latches:

$$\forall i \in L, c \in C \qquad D_i^c = \max(I_{c,\, p_i}, A_i^c) \qquad (6.17)$$

These departure constraints now capture the nonnegativity constraint of the latch.

Propagation Constraints for All Elements:

$$\forall i \in L \cup F, j \in L \cup F, c \in C \qquad A_i^c \geq D_j^c + \Delta_{DQj} + \Delta_{ji} + S_{p_j p_i} \qquad (6.18)$$

Note that the propagation constraint uses Δ_{DQ} for flip-flops, which we will define to be equal to Δ_{CQ} for notational convenience.

Although this formulation has more variables than the formulations including only latches, it actually involves less computation: the arrival times of latches are just intermediate variables requiring no more computation, and flip-flop analysis is simpler than latch analysis because time borrowing never occurs. Also note that we can use the same setup constraints for flip-flops as for latches if we substitute $T_{p_i} = 0$ for flip-flop clocks.

6.3.2   Domino Gates

Domino gates can easily be extended to this framework. When inputs to a domino gate are monotonically rising, they may arrive after the gate has entered evaluation and the domino gate may be modeled exactly as a latch. When the inputs to the domino gate are not monotonically rising, they must arrive before the gate has entered evaluation and the gate may be modeled as a flip-flop for cycle time calculations, with the additional caveat that the inputs must not change while the gate is evaluating; that is, the hold time is quite long. Hold times only appear in min-delay calculations and are discussed in the next section. Additional constraints can be added to ensure precharge finishes in time. The amount of skew budgeted at the interface of nonmonotonic to domino logic depends on the skew between launching and receiving clocks. In the case of a path that starts at a domino gate, passes through some nonmonotonic logic, and loops back to the same domino gate, the skew may be only the cycle-to-cycle jitter of the domino clock. In summary, we can determine the monotonicity of inputs from the timing type labels described in Section 4.1.4 and model domino gates as latches or flip-flops, accordingly, with additional constraints to ensure precharge is fast enough.

6.4   Min-Delay

Timing analyzers must compute not only long paths, but also short paths. Indeed, short paths are more serious because a chip can operate at reduced clock frequency if paths are longer than predicted, but will not operate at any frequency if min-delay constraints are not met. Such min-delay analysis checks that data launched from one latch or flip-flop will not propagate through logic so quickly as to violate the hold time of the next clocked element. Therefore, min-delay analysis must check only from one element to its successor; this is much easier than cycle time analysis in which a path may borrow time through many transparent latches.

To avoid min-delay failure, data departing one element must pass through enough delay that it does not violate the hold time of the next element. The earliest that data could possibly depart an element is at time 0 with respect to the element's local clock; this earliest time is guaranteed to occur if the chip is run at reduced frequency where no time borrowing occurs. We define minimum propagation delays through the clocked element and combinational logic:

- Δ_{CDi}: the hold time for latch i required between the trailing edge of the clock input and the time data changes again

- δ_{DQi}: the minimum propagation delay of latch i from the data input to the data output while the clock input is high

- δ_{ij}: the minimum propagation delay through combinational logic between latch i and latch j. If there are no combinational paths from latch i to latch j, $\delta_{ij} \equiv \infty$.

Equation 6.19 describes this min-delay constraint between adjacent latches and flip-flops. A circuit is safe from race-through if, for every consecutive pair of clocked elements, data from the earlier element cannot arrive at the later element until some hold time after the previous sampling edge of the later element. In the worst case, data departs one element on the rising edge of its clock at time 0 and arrives at the next after the minimum propagation delay through the element and combinational logic: $\delta_{DQj} + \delta_{ji}$. Time is adjusted by the phase shift operator to be relative to the receiver's clock: $S_{p_j p_i}$. Data must not arrive at the receiver until a hold time Δ_{CDi} after its sampling edge T_{p_i} of the previous cycle $- T_c$; clock skew $t_{\text{skew}}^{p_j\,(r),\,p_i(r/f)}$ between the launching and receiving clocks effectively increases the hold time. Note that the sampling edge is the falling edge for latches, but the rising edge for flip-flops. As in Section 6.3, we substitute $T_i = 0$ for edge-triggered flip-flops.

$$\forall i, j \in L \cup F \qquad \delta_{DQj} + \delta_{ji} + S_{p_j p_i} \geq T_{p_i} + \Delta_{CDi} + t_{\text{skew}}^{p_j(r),\,p_i(r/f)} - T_c \qquad (6.19)$$

Better estimates of the skew between launching and receiving clocks makes guaranteeing min-delay constraints easier. A conservative design may assume a single worst-case skew between all elements on the chip; this leads to excessive minimum propagation delay requirements between elements. By using a skew hierarchy or computing actual skews between each clock, smaller skews can be budgeted between nearby elements.

As discussed in Section 5.1.2, excess skew causes complete failure in the case of hold time violations, but only reduced operating frequency in setup time violations. Therefore, the designer may use a more conservative skew budget for min-delay than for max-delay analysis. Fortunately, min-delay races occur between clocks launched from the same global clock edge on the same cycle, so the skew budget does not include cycle-to-cycle jitter or duty cycle variation.

6.5 A Verification Algorithm

Szymanski and Shenoy present a relaxation algorithm for verifying that timing constraints are met at a given cycle time assuming no clock skew [85]. We extend the algorithm to handle arbitrary skews between elements and prune unnecessary constraints, as shown in the pseudocode of Figure 6.4. This section is rather detailed and of primary interest to those building efficient static timing analyzers, so you may skip it if you like. A key aspect of the algorithm is the introduction of extra variables for each latch, D_i^{max} and c_i^{max}, which track the latest departure from a latch with respect to any launching clock so that other paths through the latch can be pruned if they cannot be critical.

Let us first see how this algorithm handles latches, and then return to the simpler case of flip-flops. The algorithm initializes the departure times from each latch with respect to its own clock to be zero. It also initializes a variable D_i^{max} to track the latest departure time from the latch with respect to any clock and a variable c_i^{max} to track the clock that launched that latest departure (Step 3). The algorithm then follows paths from each latch to its successors and computes the arrival time at the successors with respect to the launching clock (Step 11).

A key idea of the algorithm is to prune paths that arrive early enough that they could not possibly be more critical than existing paths. To be potentially more critical and hence avoid pruning, an arrival time must satisfy two conditions (Step 12). One is that the arrival time must be later than all other departure times with respect to the same clock. The other is that the arrival time must potentially be as critical as the latest previously discovered departure time. If there were no clock skew or a single global skew budget everywhere, an arrival would only be more critical than the latest existing departure time if it actually were later: $A > D_i^{max}$. However, we allow different amounts of skew between different clocks. Figure 6.5 shows how this complicates our pruning.

Suppose that $t_{skew}^{\phi_{1a}, \phi_{1c}} = 3$, while $t_{skew}^{\phi_{1b}, \phi_{1c}} = 1$. Suppose that the departure time from L_3 D_3^{1b} on a path launched from L_2 is 2 units and that we find a path arrives at L_3 from L_1 at time 1 unit. Can we prune this path? If the clock skews were the same, we could because the path from L_1 arrives earlier than the path from L_2. Because the clock skews are different, however, data launched from L_1 must arrive at L_4 earlier than data launched

```
1 For each latch i                              ; Initialization
2              D_i^{p_i} = 0
3              D_i^{max} = 0;  c_i^{max} = p_i
4              Enqueue D_i^{p_i}
5 For each flip-flop i
6              D_i^{p_i} = 0
7              Enqueue D_i^{p_i}
8 While queue is not empty                       ; Iteration
9              Dequeue D_j^c
10             For each latch i in fanout of j
11                      A = D_j^c + Δ_{DQj} + Δ_{ji} + S_{p_j p_i}
12                      If (A > D_i^c) AND (A + t_{diff}^{c_i^{max}(r), c(r)} > D_i^{max})
13                              If (A + Δ_{DCi} + t_{skew}^{c(r), p_i(f)} > T_{p_i})
14                                      Report setup time violation
15                              Else
16                                      D_i^c = A
17                                      Enqueue D_i^c
18                                      If (A > D_i^{max})
19                                              D_i^{max} = A;  c_i^{max} = c
20             For each flip-flop i in fanout of j
21                      A = D_j^c + Δ_{DQj} + Δ_{ji} + S_{p_j p_i}
22                      If (A + Δ_{DCi} + t_{skew}^{c(r), p_i(r)} > 0)
23                              Report setup time violation
```

Figure 6.4   Pseudocode for timing verification

from L_2. Therefore, the path from L_1 may also be critical even though its departure time is earlier.

Clearly, if one path arrives at a latch more than the worst-case global clock skew before another, the early path cannot possibly be critical and may be trimmed. We can prune more aggressively by computing the dif-

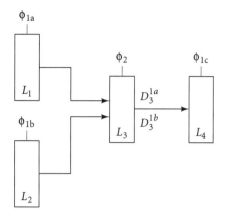

Figure 6.5   Pruning of paths with different clock skews

ference in clock skews between a pair of launching clocks ϕ_i and ϕ_j and any possible receiving clock, $t_{\text{diff}}^{\phi_i, \phi_j}$:

$$\forall \phi_r \in C \quad t_{\text{diff}}^{\phi_i, \phi_j} = \max(t_{\text{skew}}^{\phi_i, \phi_r} - t_{\text{skew}}^{\phi_j, \phi_r}) \qquad (6.20)$$

From this definition, we can show that $t_{\text{diff}}^{\phi_i, \phi_j} \leq t_{\text{skew}}^{\phi_i, \phi_j}$. Moreover, in a system budgeting a single global skew between all clocks, $t_{\text{diff}} = 0$ and negative departure times never occur, agreeing with the single skew formulation.

We check this criteria (Step 12), pruning all paths with arrival times before the latest departure by more than the clock skew between the launching clock c of the path under consideration and the launching clock $c_i^{\max}$ of the path causing the latest departure time. If the path is not pruned, it is checked for setup time violations and added to the queue so that paths to subsequent latches can be checked. Also, if it is later than the latest previously discovered departure time, it replaces the previous time (Step 19).

Flip-flops are handled in a similar fashion, but are much simpler because no time borrowing takes place. As discussed in Section 6.3.2, domino gates are analyzed either as latches or flip-flops, depending on the monotonicity of the inputs.

The algorithm performs a depth-first path search if elements are enqueued at the head of the queue and a breadth-first search if elements are enqueued at the tail. Breadth-first is likely to be faster because it can prune paths earlier.

The algorithm is very similar to one that assumes no clock skew, but may take longer because it may trace multiple paths through the same latch. This occurs when paths originating at different latches with skew between them all arrive at a common latch at nearly the same time. Fortunately real systems tend to have a relatively small number of critical paths passing through any given latch so the runtime is likely to increase by much less than the number of constraints. Timing analysis with clock domains is similar to analysis with exact skew. The runtime may be somewhat improved because a hierarchy of h levels of clock domains must trace at most h paths through any given latch. Of course, the results are more conservative.

So far, we have addressed the question of verifying that a design meets a cycle time goal because this is the primary question asked by designers. It is also straightforward to compute the minimum cycle time of a design using Sakallah's linear programming approach [70]. The constraints as presented are not quite linear because they involve the max function. The max function can be replaced by multiple inequalities, transforming the constraints into linear inequalities while preserving the minimum cycle time. These inequalities can be solved by linear programming techniques. Although conventional linear programming is much slower than the relaxation algorithm for verifying cycle time, new interior point methods [95] may be quite efficient.

6.6   Case Study

To evaluate the costs and benefits of the exact formulation, we analyzed a timing model of MAGIC, the Memory and General Interconnect Controller of the FLASH supercomputer [50], implemented in a 0.6-micron CMOS process. MAGIC includes a two-way superscalar RISC processing engine and several large data buffers. We extracted a timing model from the Standard Delay Format (SDF) data produced by LSI Logic tools, then trimmed long paths such as those involving reset or scan. After trimming, we found 1819 latches and 10,559 flip-flops connected by 593,153 combinational paths (Model A). To obtain an entirely latch-based design, we replaced each flip-flop with a pair of latches and divided the path delay between the two latches, obtaining a system with 22,937 latches (Model

B). The chip was partitioned into 10 units, each a local clock domain. We assumed 500 ps of global skew between domains and 250 ps of local skew within domains.

We applied the timing analysis algorithm of Section 6.5 to the timing model. Table 6.2 shows the minimum cycle times achievable and number of latch departures enqueued in each run, a measure of the analysis cost. Model B is uniformly faster than Model A because latches allow the system to borrow time across cycles, solving some critical paths. The exact analysis shows that the system can run 50–90 ps faster than a single skew analysis conservatively predicts. Each latch departure is enqueued at least once when its departure time is initialized to 0. Paths borrowing time enqueue later departure times. The exact analysis also enqueues more latch departures because potentially critical paths from multiple launching clocks may pass through a single latch. The exact analysis enqueues 143 more than the single skew analysis in Model A and 333 more in Model B. These differences are less than 4% of the total number of departures, indicating that pruning makes the exact analysis only slightly more expensive than the single skew approximation. In all cases, the CPU time for analysis is under a second, much shorter than the time required to read the timing model from disk.

These results indicate that the exact skew formulation works well in practice because only a small fraction of paths require time borrowing, as noted by Szymanski and Shenoy [85], and because an even smaller fraction of paths involve negative departure times. In this particular problem, no critical paths depart a clock domain and return to it, so the clock domain formulation would have found equally good cycle times. However, the cost of the exact skew formulation is low enough that no approximations are necessary.

Table 6.2 Timing analysis results

	Model A	Model B
Single skew	9.43 ns	8.05 ns
	3866 departures	24,995 departures
Exact skew	9.38 ns	7.96 ns
	4009 departures	25,328 departures

6.7   Historical Perspective

Early efforts in timing analysis, surveyed in [37], only considered edge-triggered flip-flops. Thus they had to analyze just the combinational logic blocks between registers because the cycle time is set by the longest combinational path between registers. Netlist-level timing analyzers, such as CRYSTAL [63] and TV [45], used switch-level RC models [69] to compute delay through the combinational blocks.

Many circuits use level-sensitive latches instead of flip-flops. Latches complicate the analysis because they allow time borrowing: a signal that reaches the latch input while the latch is transparent does not have to wait for a clock edge, but rather can immediately propagate through the latch and be used in the next phase of logic. Analysis of systems with latches was long considered a difficult problem [63], and various netlist-level timing analyzers applied heuristics for latch timing, but eventually Unger [89] developed a complete set of timing constraints for two-phase clocking with level-sensitive latches. LEADOUT [83], by Szymanski, checked timing equations to properly handle multiphase clocking and level-sensitive latches. Champernowne et al. [8] developed a set of latch-to-latch timing rules that allow a hierarchy of clock skews but do not permit time borrowing.

Sakallah et al. [70] provide a very elegant formulation of the timing constraints for latch-based systems. They show that maximum-delay constraints can be expressed with a system of inequalities. They then use a linear programming algorithm to minimize the cycle time and to determine an optimal clock schedule. Because the clock schedule is usually fixed and the user is interested in verifying that the circuits can operate at a target frequency, more efficient algorithms can be used to process the constraints, such as the relaxation approach suggested by Szymanski and Shenoy [85]. Moreover, many of the constraints in the formulation may be redundant, so graph-based techniques proposed by Szymanski [84] can determine the relevant constraints. Ishii et al. [44] offer yet another efficient algorithm for verifying the cycle time of two-phase latched systems. Burks et al. [7] express timing analysis in terms of critical paths and support clock skew in limited ways.

Little published research has taken place in timing analysis since the early 1990s. Instead, commercial static timing analyzers have reached maturity. Tools such as Pathmill and Pearl are powerful and efficient enough to handle multimillion transistor designs using aggressive circuit techniques. Indeed, at the time of this writing, Pathmill is adding a "Clock Skew Option" that supports different amounts of clock skew between different clocks as described in this chapter. PrimeTime, also from Synopsys, allows the user to enter different amounts of clock skew between different clocks. Unfortunately the tool is pessimistic because it budgets the skew between the latches at each end of each half-cycle rather than the skew between the launching and receiving clocks. For example, in Figure 6.2, on a path starting at L_5, passing through L_6 and L_7 while they are transparent, and finally arriving at L_4, the tool budgets the global skew between L_7 and L_4 rather than the local skew from L_5 to L_4. The moral is to scrutinize carefully claims from CAD marketing departments about how clock skew is handled.

6.8 Summary

In this chapter, we have extended the latch-based timing analysis formulation of Sakallah et al. to handle clock skew, especially different amounts of clock skew between different elements. Allowing a single amount of clock skew everywhere effectively increases the setup time of each latch. An exact analysis allowing different amounts of skew between different elements involves tracking the clock that launched each path so that paths that leave a local skew domain and then return only budget the local skew. This leads to a multiplication of constraints proportional to the number of clocks, but most constraints are not tight. Most practical systems use clocked elements besides just transparent or pulsed latches, so we also incorporate edge-triggered flip-flops and domino gates into the timing analysis formulation by separately tracking arrival and departure times at each clocked element. In addition to verifying cycle time, we check for min-delay violations, effectively increasing the hold time of each element by the potential clock skew between launching and receiving elements. Finally, we presented a relaxation algorithm for verifying the timing constraints. The uncertainty from the clock skew may increase the number of paths that must be searched, but the case study of the MAGIC chip shows that this increase is very modest because most paths do not borrow time.

6.9   Exercises

[15]   **6.1**   Consider the path in Figure 6.6 using flip-flops F_1 and F_2. The flip-flops have a setup time of 0.2 ns and a clock-to-Q delay of 0.3 ns. There is no skew between the clocks ϕ_{1a} and ϕ_{1b} that share the same start time. The departure time from each flop is $D_1 = D_2 = 0$ by definition.

(a) What are the arrival times A_1 and A_2?

(b) What is the minimum cycle time at which the system operates correctly?

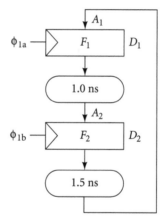

Figure 6.6   Path for Exercise 6.1

[15]   **6.2**   Suppose you were free to choose the start time of ϕ_{1b} independent of ϕ_{1a} in Exercise 6.1. What start time would you choose to minimize the cycle time? What cycle time could be achieved?

[15]   **6.3**   Repeat Exercise 6.1 if a skew of 100 ps is budgeted between ϕ_{1a} and ϕ_{1b}.

[15]   **6.4**   Repeat Exercise 6.2 if a skew of 100 ps is budgeted between ϕ_{1a} and ϕ_{1b}.

[15]

6.5   Consider the path in Figure 6.7 using flip-flops F_1, F_2, and F_3. The flip-flops have a setup time of 0.1 ns and a clock-to-Q delay of 0.15 ns.

There is no skew between the clocks ϕ_{1a}, ϕ_{1b}, and ϕ_{1c} that all share the same start time. The departure time from each flop is $D_1 = D_2 = D_3 = 0$ by definition.

(a) What are the arrival times A_1, A_2, and A_3?

(b) What is the minimum cycle time at which the system operates correctly?

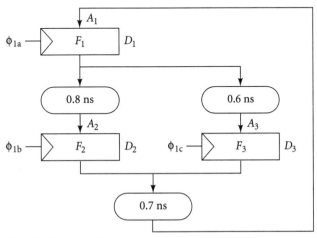

Figure 6.7   Path for Exercise 6.5

[20]   **6.6**   Referring to the information in Exercise 6.5, suppose ϕ_{1a} and ϕ_{1b} are in a common local clock domain, but ϕ_{1c} is in a different clock domain. What is the minimum cycle time of the system if

(a) the local skew is 50 ps and the global skew is 140 ps?

(b) the local skew is 25 ps and the global skew is 300 ps?

[25]   **6.7**   Consider the microprocessor core from Example 6.1. For each of the following sets of logic delays, determine if the system will function at 100 MHz with 50% duty cycle clocks, assuming no clock skew. If so, calculate the arrival and departure times A_4, A_5, A_6, A_7, D_4, D_5, D_6, and D_7, and determine which latches experience time borrowing, if any.

(a) $\Delta 4 = 3$ ns; $\Delta 5 = 4$ ns; $\Delta 6 = 2$ ns; $\Delta 7 = 5$ ns; $\Delta_{DQ} = 0$; $\Delta_{DC} = 0$

(b) $\Delta 4 = 3$ ns; $\Delta 5 = 6$ ns; $\Delta 6 = 7$ ns; $\Delta 7 = 1$ ns; $\Delta_{DQ} = 0$; $\Delta_{DC} = 0$

(c) $\Delta 4 = 4$ ns; $\Delta 5 = 7$ ns; $\Delta 6 = 3$ ns; $\Delta 7 = 2$ ns; $\Delta_{DQ} = 0$; $\Delta_{DC} = 0$

(d) $\Delta 4 = 3$ ns; $\Delta 5 = 5$ ns; $\Delta 6 = 2$ ns; $\Delta 7 = 3$ ns; $\Delta_{DQ} = 1$ ns;
$\Delta_{DC} = 1$ ns

(e) $\Delta 4 = 1$ ns; $\Delta 5 = 8$ ns; $\Delta 6 = 2$ ns; $\Delta 7 = 3$ ns; $\Delta_{DQ} = 0.1$ ns;
$\Delta_{DC} = 3$ ns

[25] **6.8** Consider the microprocessor core from Example 6.1. For each of the
following sets of logic delays, determine if the system will function at 1
GHz with 50% duty cycle clocks, assuming 200 ps of clock skew between
any pair of clocks. If so, calculate the arrival and departure times A_4, A_5,
A_6, A_7, D_4, D_5, D_6, and D_7, and determine which latches experience time
borrowing, if any.

(a) $\Delta 4 = 500$ ps; $\Delta 5 = 500$ ps; $\Delta 6 = 400$ ps; $\Delta 7 = 600$ ps; $\Delta_{DQ} = 0$;
$\Delta_{DC} = 0$

(b) $\Delta 4 = 850$ ps; $\Delta 5 = 100$ ps; $\Delta 6 = 400$ ps; $\Delta 7 = 400$ ps; $\Delta_{DQ} = 0$;
$\Delta_{DC} = 50$ ps

(c) $\Delta 4 = 200$ ps; $\Delta 5 = 500$ ps; $\Delta 6 = 500$ ps; $\Delta 7 = 200$ ps;
$\Delta_{DQ} = 100$ ps; $\Delta_{DC} = 150$ ps

(d) $\Delta 4 = 400$ ps; $\Delta 5 = 300$ ps; $\Delta 6 = 500$ ps; $\Delta 7 = 350$ ps;
$\Delta_{DQ} = 100$ ps; $\Delta_{DC} = 150$ ps

[35] **6.9** Repeat the example from Section 6.2.4, computing the minimum T_c
under the exact, clock domain, and single skew formulations for each of
the following combinational logic delays. When T_c is greater for the clock
domain or single skew formulation than the exact formulation, explain
why the simpler formulation is pessimistic.

(a) $\Delta 4 = 6$; $\Delta 5 = 6$; $\Delta 6 = 6$; $\Delta 7 = 6$

(b) $\Delta 4 = 7$; $\Delta 5 = 5$; $\Delta 6 = 3$; $\Delta 7 = 5$

(c) $\Delta 4 = 7$; $\Delta 5 = 3$; $\Delta 6 = 5$; $\Delta 7 = 4$

(d) $\Delta 4 = 0.5$; $\Delta 5 = 11.5$; $\Delta 6 = 5$; $\Delta 7 = 6$

(e) $\Delta 4 = 1.5$; $\Delta 5 = 10.5$; $\Delta 6 = 6$; $\Delta 7 = 6$

(f) $\Delta 4 = 10$; $\Delta 5 = 2$; $\Delta 6 = 6$; $\Delta 7 = 6$

(g) $\Delta 4 = 10$; $\Delta 5 = 2$; $\Delta 6 = 5$; $\Delta 7 = 7$

(h) $\Delta 4 = 8$; $\Delta 5 = 3$; $\Delta 6 = 7$; $\Delta 7 = 6$

(i) $\Delta 4 = 8$; $\Delta 5 = 2$; $\Delta 6 = 7$; $\Delta 7 = 7$

[35]   **6.10** Use the Solver in the Excel spreadsheet or another linear programming package to calculate the minimum T_c for each of the cases in Exercise 6.9.

[20]   **6.11** Consider the path in Figure 6.8 using flip-flops F_1 and F_2. Clocks ϕ_{1a}
and ϕ_{1b} nominally share the same start time if there is no skew.

   (a) If each flip-flop has a hold time of 100 ps and a contamination
       delay of 50 ps, what is the minimum value of $\delta1$ for correct
       operation if there is no clock skew?

   (b) Repeat part (a) if the clock skew is budgeted at 75 ps.

   (c) Suppose there is no logic to perform between F_1 and F_2. How
       could the designer ensure the hold time is still met at F_2 in the
       scenario from part (b)?

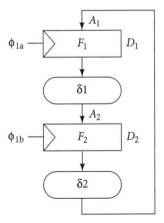

Figure 6.8   Path for Exercise 6.11

[15]   **6.12** Consider the path in Figure 6.9 using flip-flops F_1, F_2, and F_3. Suppose the hold times of F_1 and F_3 are 60 ps and of F_2 is 30 ps. Let the contamination delay of each flop be 90 ps. If ϕ_{1a} and ϕ_{1b} are in a local clock
domain experiencing 50 ps of skew and ϕ_{1c} is in a separate clock domain
experiencing up to 150 ps of skew relative to the other clocks, what are
the minimum contamination delays $\delta1$, $\delta2$, and $\delta3$ to ensure correct
operation?

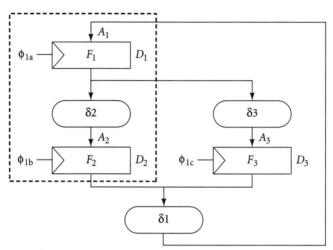

Figure 6.9   Path for Exercise 6.12

[15]   **6.13** Consider the microprocessor core from Example 6.1. Let each latch have a hold time of 50 ps and a contamination delay of 60 ps.

(a) If there is no skew, what is the minimum contamination delay through each block of logic for correct operation?

(b) If there is 150 ps of skew between clocks, what is the minimum contamination delay through each block of logic for correct operation?

[20]   **6.14** Consider the microprocessor core from Figure 6.2. Let each latch have a hold time of 40 ps and a contamination delay of 30 ps. If the skew between clock domains is 100 ps and the skew within clock domains is 60 ps, what are the minimum contamination delays, $\delta 4$, $\delta 5$, $\delta 6$, and $\delta 7$, to ensure correct operation?

7
Conclusions

And so our victorious heros return Princess Clock to the happy Land of Giga Chips. The vanquished Dragon of Skew will never threaten chip designers again...

Prediction is very difficult, especially if it's about the future.
—Nils Bohr

As cycle times in high-performance digital systems shrink faster than mere process improvement allows, sequencing overhead consumes an increasing fraction of the clock period. Flip-flops and traditional domino circuits, in particular, suffer from clock skew, latch delay, and the inability to balance logic between cycles through time borrowing. The overhead of traditional domino circuits can waste 25% or more of the cycle time in aggressive systems! Fortunately, the designer can hide much of this overhead through better design techniques. Static pipelines built from transparent latches can tolerate nearly half a cycle of clock skew and help the designer balance logic with time borrowing. Pulsed latches offer similar advantages, trading some skew tolerance and time borrowing for a faster latch. Skew-tolerant domino circuits are particularly fast, completely eliminating latch delay and tolerating modest amounts of clock skew and time borrowing. Smaller amounts of skew can be budgeted in local clock domains than across a large die, reducing the burden of clock skew.

Chapter 2 explored the design of static circuits using flip-flops, transparent latches, and pulsed latches. We found that the purpose of such elements is not so much to remember information as to sequence information along a pipeline or through a state machine, preventing data in one stage from interfering with data in another. The elements must slow down fast paths to prevent interference while minimizing extra delay on paths that are already critical. Because it is impossible to slow some paths without at least slightly impeding all others, these elements inevitably introduce sequencing overhead. The sequencing overhead imposed by the hard edges of flip-flops is worst, including two latch delays and clock skew. Transparent latches are faster, hiding the clock skew. Pulsed latches can be even faster, introducing only one latch delay while still possibly hiding the clock skew. The speed of pulsed latches comes at the expense of longer hold times. Transparent latches and pulsed latches are also good because they provide a window during which data may arrive without extra delay. In addition to hiding clock skew, this window allows logic to borrow time across cycles to balance logic. Pulsed latches and flip-flops have frequently been mixed up in the literature because they both are used once per cycle; pulsed latches can be distinguished by their window of transparency.

Chapter 3 moved on to the design of domino circuits. Domino circuits offer raw gate delays 1.5 to 2 times faster than static circuits, making them very popular for high-speed designs. Unfortunately, traditional domino design techniques also impose hard edges at every half-cycle boundary, leading to enormous overhead of two latch delays and twice the clock skew in every cycle! Skew-tolerant domino circuits use multiple overlapping clock phases and eliminate latches to soften these hard edges, removing the sequencing overhead entirely.

Chapter 4 united skew-tolerant domino with transparent latches and pulsed latches in a systematic four-phase skew-tolerant circuit design methodology. The interface from static to domino logic inherently must budget clock skew, motivating the designer to build entire critical loops from dual-rail domino circuits to avoid this penalty. Skew-tolerant domino also integrates seamlessly with RAMs, PLAs, and other dynamic structures. With four clock phases and many different types of clocked elements, it is easy to become confused about legal connections. By tagging each signal with a timing type, it is simple to verify connectivity. The methodology also includes scanning data in and out of both static and domino pipeline stages to help testability.

None of these skew-tolerant circuit techniques would be useful if the clock generators were too complex or introduced more skew than they tolerated. Chapter 5 examined clocking, beginning with the often used but seldom defined term *clock skew*. The designer wishes to receive a small number of logical clocks with precisely defined phase relationships arriving at all parts of the chip simultaneously. Variations in the global and local clock generation and distribution circuits cause the designer to actually receive slightly different physical clocks at each point. These variations can be categorized as predictable or unpredictable and as DC, slowly varying, or rapidly varying; different techniques can be used to handle different components. Ultimately, it is very difficult to reduce worst-case clock skew below 200 ps across a complex chip. When such skews become a significant problem, the designer can introduce clock domains to budget smaller amounts of skew between local elements than across the entire chip. Unfortunately, clock domains do not reduce duty cycle variation, which is an increasingly important component of skew.

Design techniques are of little value unless accompanied by suitable verification tools. In particular, most static timing analyzers from the mid1990s are unable to take advantage of reduced clock skews in local clock domains. Chapter 6 addressed timing analysis, showing that arrival times cease to have absolute meaning in systems with different skews between different elements. Instead, arrival times must be specified with reference to a particular launching clock that determines the skew relative to the receiver. Therefore, timing analysis introduces a vector of arrival times at each latch with respect to different launching clocks. Fortunately, this vector is relatively sparse because most paths do not borrow time in a real system.

In summary, conventional designs with flip-flops and traditional domino clocking are becoming inadequate for high-speed designs. Systems operating above 1 GHz will be unlikely to achieve acceptably low global skew across the entire die at reasonable cost. Instead of abandoning the synchronous paradigm entirely for an asynchronous design, designers will divide the die into local clock domains offering smaller amounts of skew within each domain and will use skew-tolerant circuit design techniques to hide this modest amount of skew. Transparent latches have a long history of successful use; pulsed latches bring larger min-delay constraints, but are even faster and have been successfully used on large microprocessors. Skew-tolerant domino can achieve zero overhead, offering the full speedup of domino gates. With such approaches, we expect clocked systems will remain viable to extremely high operating frequencies.

As systems grow to include hundreds of millions of transistors operating at many gigahertz, circuit designers will encounter even more challenges. Although skew-tolerant techniques can potentially hide up to half a cycle of clock skew, design becomes very difficult at such extremes. If global skew does not fall well below 200 ps, standard approaches to global communication will not work above 2.5 GHz. Communication between different clock domains may have to occur at reduced frequency or via an asynchronous interface [10, 23]. Even within a local clock domain, duty cycle variation will cut into the amount of time available for borrowing and may eventually require local correction. Moreover, domino circuits face an impending power crunch. Chip performance will become power limited because only a finite amount of heat can be removed from a die with a reasonably priced cooling system. Although dual-rail domino gates are extremely fast, they are much more power hungry than static circuits because their activity factors are usually far greater. Improvements in clock gating will disable some inactive domino gates, but domino will continue to pay a large power premium. Will domino gates become too expensive from a power perspective, or will designers find it better to build simpler machines with fewer transistors running at extreme domino speeds than very complex machines churning along at the speed of static logic? Domino presents other challenges as well. The aspect ratios of wires continue to grow, making coupling problems greater. Scaling device thresholds increase leakage currents and reduce noise margins. The design time of domino also can be high, possibly increasing time to market. Will static circuits become a better choice for teams with finite resources, or will advances in CAD tools improve domino productivity? Circuit design should remain an exciting field as these issues are explored in the coming decade.

A

Timing Constraints

To illustrate the formulations described in Sections 6.1 and 6.2, we present a complete set of the timing constraints of each formulation applied to the simple microprocessor example from Figures 6.1 and 6.2.

Skewless Formulation

The timing constraints with no skew are the following:

Setup Constraints

$$D_4 \leq T_{\phi_1} \qquad D_5 \leq T_{\phi_2} \qquad D_6 \leq T_{\phi_1} \qquad D_7 \leq T_{\phi_2}$$

Propagation Constraints

D_4 = max of	D_5 = max of	D_6 = max of	D_7 = max of
0	0	0	0
$D_3 + \Delta 4 - T_p$	$D_4 + \Delta 5 - T_p$	$D_5 + \Delta 6 - T_p$	$D_6 + \Delta 7 - T_p$
$D_5 + \Delta 4 - T_p$			
$D_7 + \Delta 4 - T_p$			

Single Skew Formulation

The timing constraints budgeting global skew everywhere are very similar to those with no skew:

Setup Constraints

$$D_4 + T_{\text{skew}}^{\text{global}} \leq T_{\phi_{1a}} \qquad D_5 + T_{\text{skew}}^{\text{global}} \leq T_{\phi_{2a}} \qquad D_6 + T_{\text{skew}}^{\text{global}} \leq T_{\phi_{1b}} \qquad D_7 + T_{\text{skew}}^{\text{global}} \leq T_{\phi_{2b}}$$

Propagation Constraints

D_4 = max of	D_5 = max of	D_6 = max of	D_7 = max of
0	0	0	0
$D_3 + \Delta 4 - T_p$	$D_4 + \Delta 5 - T_p$	$D_5 + \Delta 6 - T_p$	$D_6 + \Delta 7 - T_p$
$D_5 + \Delta 4 - T_p$			
$D_7 + \Delta 4 - T_p$			

Exact Formulation

Because there are four clocks, there are four times as many setup and propagation constraints for the exact analysis:

Setup Constraints

$$D_4^{\phi_{1a}} + T_{\text{skew}}^{\text{local}} \leq T_{\phi_{1a}} \qquad D_5^{\phi_{1a}} + T_{\text{skew}}^{\text{local}} \leq T_{\phi_{2a}} \qquad D_6^{\phi_{1a}} + T_{\text{skew}}^{\text{global}} \leq T_{\phi_{1b}} \qquad D_7^{\phi_{1a}} + T_{\text{skew}}^{\text{global}} \leq T_{\phi_{2b}}$$

$$D_4^{\phi_{2a}} + T_{\text{skew}}^{\text{local}} \leq T_{\phi_{1a}} \qquad D_5^{\phi_{2a}} + T_{\text{skew}}^{\text{local}} \leq T_{\phi_{2a}} \qquad D_6^{\phi_{2a}} + T_{\text{skew}}^{\text{global}} \leq T_{\phi_{1b}} \qquad D_7^{\phi_{2a}} + T_{\text{skew}}^{\text{global}} \leq T_{\phi_{2b}}$$

$$D_4^{\phi_{1b}} + T_{\text{skew}}^{\text{global}} \leq T_{\phi_{1a}} \qquad D_5^{\phi_{1b}} + T_{\text{skew}}^{\text{global}} \leq T_{\phi_{2a}} \qquad D_6^{\phi_{1b}} + T_{\text{skew}}^{\text{local}} \leq T_{\phi_{1b}} \qquad D_7^{\phi_{1b}} + T_{\text{skew}}^{\text{local}} \leq T_{\phi_{2b}}$$

$$D_4^{\phi_{2b}} + T_{\text{skew}}^{\text{global}} \leq T_{\phi_{1a}} \qquad D_5^{\phi_{2b}} + T_{\text{skew}}^{\text{global}} \leq T_{\phi_{2a}} \qquad D_6^{\phi_{2b}} + T_{\text{skew}}^{\text{local}} \leq T_{\phi_{1b}} \qquad D_7^{\phi_{2b}} + T_{\text{skew}}^{\text{local}} \leq T_{\phi_{2b}}$$

Propagation Constraints

$D_4^{\phi_{1a}} = \text{max of}$	$D_5^{\phi_{1a}} = \text{max of}$	$D_6^{\phi_{1a}} = \text{max of}$	$D_7^{\phi_{1a}} = \text{max of}$
0			
$D_3^{\phi_{1a}} + \Delta 4 - T_p$	$D_4^{\phi_{1a}} + \Delta 5 - T_p$	$D_5^{\phi_{1a}} + \Delta 6 - T_p$	$D_6^{\phi_{1a}} + \Delta 7 - T_p$
$D_5^{\phi_{1a}} + \Delta 4 - T_p$			
$D_7^{\phi_{1a}} + \Delta 4 - T_p$			

$D_4^{\phi_{2a}} = \text{max of}$	$D_5^{\phi_{2a}} = \text{max of}$	$D_6^{\phi_{2a}} = \text{max of}$	$D_7^{\phi_{2a}} = \text{max of}$
	0		
$D_3^{\phi_{2a}} + \Delta 4 - T_p$	$D_4^{\phi_{2a}} + \Delta 5 - T_p$	$D_5^{\phi_{2a}} + \Delta 6 - T_p$	$D_6^{\phi_{2a}} + \Delta 7 - T_p$
$D_5^{\phi_{2a}} + \Delta 4 - T_p$			
$D_7^{\phi_{2a}} + \Delta 4 - T_p$			

$D_4^{\phi_{1b}} = \text{max of}$	$D_5^{\phi_{1b}} = \text{max of}$	$D_6^{\phi_{1b}} = \text{max of}$	$D_7^{\phi_{1b}} = \text{max of}$
		0	
$D_3^{\phi_{1b}} + \Delta 4 - T_p$	$D_4^{\phi_{1b}} + \Delta 5 - T_p$	$D_5^{\phi_{1b}} + \Delta 6 - T_p$	$D_6^{\phi_{1b}} + \Delta 7 - T_p$
$D_5^{\phi_{1b}} + \Delta 4 - T_p$			
$D_7^{\phi_{1b}} + \Delta 4 - T_p$			

$D_4^{\phi_{2b}} = \text{max of}$	$D_5^{\phi_{2b}} = \text{max of}$	$D_6^{\phi_{2b}} = \text{max of}$	$D_7^{\phi_{1a}} = \text{max of}$
			0
$D_3^{\phi_{2b}} + \Delta 4 - T_p$	$D_4^{\phi_{2b}} + \Delta 5 - T_p$	$D_5^{\phi_{2b}} + \Delta 6 - T_p$	$D_6^{\phi_{1a}} + \Delta 7 - T_p$
$D_5^{\phi_{2b}} + \Delta 4 - T_p$			
$D_7^{\phi_{2b}} + \Delta 4 - T_p$			

Clock Domain Formulation

Finally, we write the constraints with the approximation of clock domains. Because there are two levels of the clock domain hierarchy, this requires only twice as many constraints as the single skew formulation. The timing constraints are the following:

Setup Constraints

$$D_4^1 + T_{skew}^{local} \leq T_{\phi_{1a}} \qquad D_5^1 + T_{skew}^{local} \leq T_{\phi_{2a}} \qquad D_6^1 + T_{skew}^{local} \leq T_{\phi_{1b}} \qquad D_7^1 + T_{skew}^{local} \leq T_{\phi_{2b}}$$

$$D_4^2 + T_{skew}^{global} \leq T_{\phi_{1a}} \qquad D_5^2 + T_{skew}^{global} \leq T_{\phi_{2a}} \qquad D_6^2 + T_{skew}^{global} \leq T_{\phi_{1b}} \qquad D_7^2 + T_{skew}^{global} \leq T_{\phi_{2b}}$$

Propagation Constraints

$D_4^1 = $ max of	$D_5^1 = $ max of	$D_6^1 = $ max of	$D_7^1 = $ max of
0	0	0	0
$D_3^1 + \Delta 4 - T_p$	$D_4^1 + \Delta 5 - T_p$		$D_6^1 + \Delta 7 - T_p$
$D_5^1 + \Delta 4 - T_p$			

$D_4^2 = $ max of	$D_5^2 = $ max of	$D_6^2 = $ max of	$D_7^2 = $ max of
$D_7^1 + \Delta 4 - T_p$		$D_5^1 + \Delta 6 - T_p$	
$D_3^2 + \Delta 4 - T_p$	$D_4^2 + \Delta 5 - T_p$	$D_5^2 + \Delta 6 - T_p$	$D_6^2 + \Delta 7 - T_p$
$D_5^2 + \Delta 4 - T_p$			
$D_7^2 + \Delta 4 - T_p$			

B

Solutions to Even-Numbered Exercises

1.2 Overhead $= \Delta_{CQ} + \Delta_{DC} + t_{skew} = 330$ ps. This is 20% of a 600 MHz cycle and 33% of a 1 GHz cycle.

1.4 Overhead $= 2\Delta_{CQ} = 2.6$ FO4 delays $= 156$ ps. This is 9% of a 600 MHz cycle and 16% of a 1 GHz cycle. Note that clock skew does not impact the cycle time in a system built from transparent latches.

1.6 $\Delta_{logic} = T_c - (\Delta_{CQ} + \Delta_{DC} + t_{skew}) = 5$ ns $- (0.427$ ns $+ 0.018$ ns $+ 0.4$ ns$) = 4.155$ ns.

1.8 See Table B.1 $T_c = 3/N + (0.1 + 0.12 + 0.05)$. Frequency $= 1/T_c$. Latency $= N \cdot T_c$.

Table B.1 Clock frequency and computation latency

# Cycles (N)	Max frequency (MHz)	Total latency (ns)
1	305	3.27
2	564	3.54
3	787	3.81
4	980	4.08

1.12 $T_c = \Delta_{logic} + 2\Delta_{DC} + 2t_{skew} = 1380$ ps. 27% of this time is overhead.

1.14 For static CMOS, $T_c = \Delta_{\text{logic}} + 2\Delta_{DQ}$. For traditional domino, $T_c = 0.7$ $\Delta_{\text{logic}} + 2\Delta_{DC} + 2t_{\text{skew}}$. Therefore, domino is faster when $t_{\text{skew}} < 0.15$ Δ_{logic}, as shown in Figure B.1.

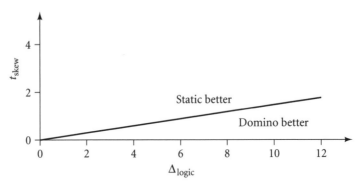

Figure B.1   Design space of logic delay and clock skew (solution)

1.16 Only b and e. The others produce Q signals that may fall while ϕ is high.

2.2 (a) From Equation 2.2, $\Delta_{\text{logic}} = 1000 \text{ ps} - 150 \text{ ps} - 90 \text{ ps} - 90 \text{ ps}$ $= 670 \text{ ps}$.

(b) From Equation 2.3, $\Delta_{\text{logic}} = 1000 \text{ ps} - 2 \cdot 70 \text{ ps} = 860 \text{ ps}$.

(c) From Equation 2.4, $\Delta_{\text{logic}} = 1000 \text{ ps} - 70 \text{ ps} - \max(0, 150 \text{ ps} + 90 \text{ ps} - t_{\text{pw}})$. For $t_{\text{pw}} = 180 \text{ ps}$, $\Delta_{\text{logic}} = 870 \text{ ps}$. For $t_{\text{pw}} = 250 \text{ ps}$, $\Delta_{\text{logic}} = 930 \text{ ps}$.

2.4 $t_{\text{nonoverlap}} = T_c(0.5 - d)$. $\Delta_1 = \Delta_2 = T_c - (\Delta_{DQ} + \Delta_{DC} + t_{\text{skew}} + t_{\text{nonoverlap}}) = 690 \text{ ps} + T_c(d - 0.5)$. $\Delta_1 + \Delta_2 = T_c - 2\Delta_{DQ} = 860 \text{ ps}$. $\delta_1 = \delta_2 = \Delta_{CD} + t_{\text{skew}} - \delta_{CQ} - t_{\text{nonoverlap}} = 130 \text{ ps} + T_c(d - 0.5)$. $\delta_1 + \delta_2 = 260 \text{ ps} + T_c(2d - 1)$.

2.6 Wide pulses hide the overhead of setup time and clock skew and may permit some time borrowing. However, wide pulses also increase min-delay constraints, making hold times more difficult to satisfy.

3.2 From Equation 3.1, $t_p = 200$ ps $+ 50$ ps $= 250$ ps.

3.4 See Figure B.2.

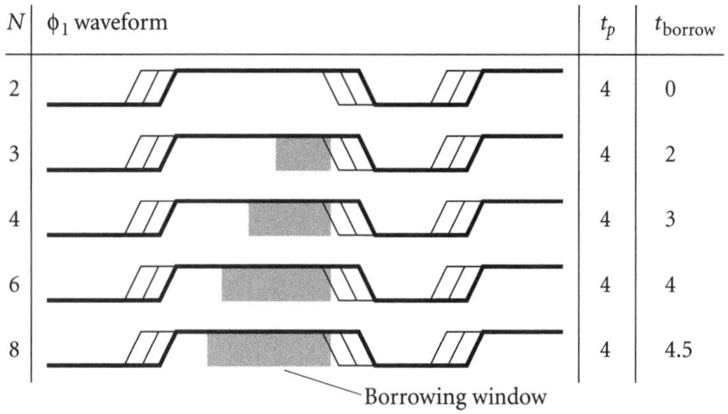

N	ϕ_1 waveform	t_p	t_{borrow}
2		4	0
3		4	2
4		4	3
6		4	4
8		4	4.5

Borrowing window

Figure B.2 Time borrowing for various numbers of clock phases

3.6 $t_{skew-max} = 475$ with $t_p = 462.5$ ps and $t_e = 787.5$ ps. When the global skew is 200 ps, $t_p = 187.5$ ps and $t_e = 1062.5$ ps, allowing 862.5 ps of time borrowing.

3.8 See Figure B.3.

3.10 See Figure B.4. The estimated logical efforts are (a) 2/3; (b) 1; (c) 1/3; (d) 1/3; (e) 2/3. The actual logical efforts are probably lower because of the lower switching threshold of a dynamic gate.

3.18 The noise margin on the D input of circuit (a) is only a threshold voltage when the input and clock are both low. If D falls a threshold voltage below the low voltage on the clock, on account of ground noise or coupling on the long connections between units, the pass gate will turn on and discharge the dynamic node. Circuit (b) has the full noise margin of the inverter and is thus much more robust.

4.2 (a) See Figure B.5. At a cycle of 1.5, time is borrowed through L_3, L_4, and L_5. At a cycle of 1, time is borrowed through L_3 and the setup time at L_4 is missed. At a cycle time of 0.8, the setup time at L_3 is missed.

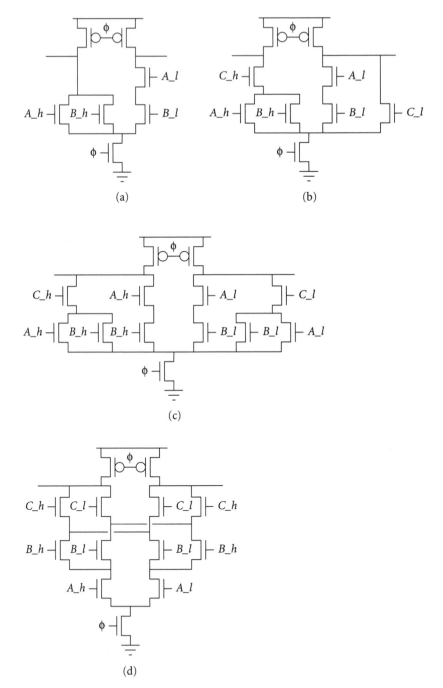

Figure B.3 Dual-rail domino gates

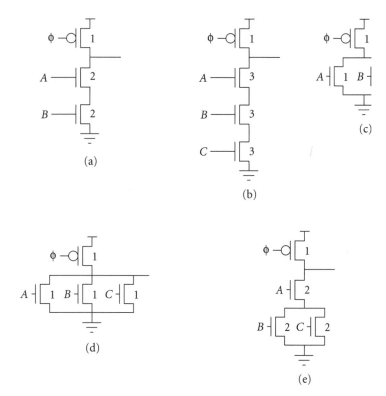

Figure B.4   Unfooted dynamic gates with transistor sizes

(b) See Figure B.6. At a cycle time of 1.5, time is borrowed through L_4, L_5, and L_2. At a cycle time of 1, time is borrowed through L_4 and the setup time at L_5 is missed. At a cycle time of 0.8, the setup time at L_4 is missed.

4.4   See Figure B.7.

5.2   The designer can predict and take advantage of systematic variations in the start time of two physical clocks by adjusting the amount of logic in each cycle to fit the actual time available. The designer must pessimistically assume random variations reduce the amount of time available for logic.

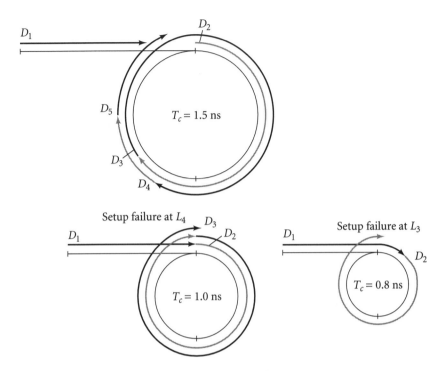

Figure B.5   Time borrowing diagram for latch-based system

5.4 The pulse stretcher ensures ϕ_6 overlaps the next cycle of ϕ_1 so that data is not lost. If it were omitted and the clock were run slowly or the delay chain was faster than nominal, gate X might precharge before gate Y consumes the result, as shown in Figure B.8. Thus, the hold time would be violated and the circuit would produce the wrong result.

5.6 Clock networks such as H-trees may have zero systematic skew if perfectly balanced, but inevitably have some random skew and jitter because of process, voltage, and temperature differences across the die. Therefore, the designer must still budget some skew.

6.2 $T_c = 1.75$ ns using $T_{\phi 1b} = -0.25$ ns modulo $T_c = 1.5$ ns.

6.4 $T_c = 1.85$ ns using $T_{\phi 1b} = -0.25$ ns modulo $T_c = 1.6$ ns.

6.6 (a) $T_c = 1.10$ ns; (b) $T_c = 1.25$ ns.

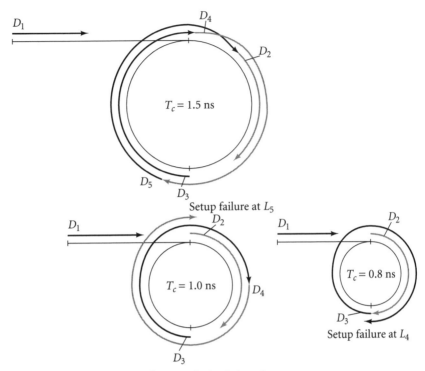

Figure B.6   Time borrowing diagram for latch-based system

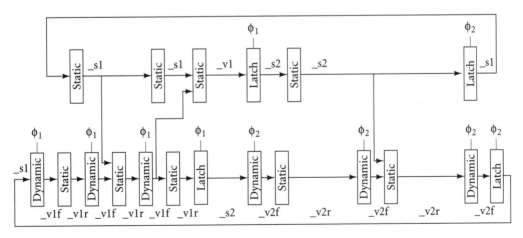

Figure B.7   Timing types for Exercise 4.4

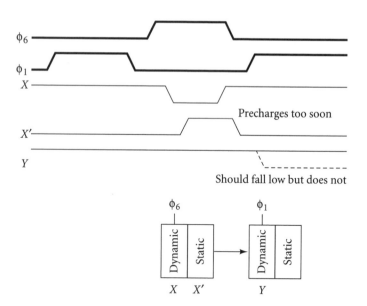

Figure B.8   Circuit failure without pulse stretcher

6.8   See Table B.2.

Table B.2   Latch arrival and departure times (ps)

	Operates	A_4	A_5	A_6	A_7	D_4	D_5	D_6	D_7	Borrowing
a	Yes	100	100	0	100	100	100	0	100	L_4, L_5, L_7
b	Yes	350	−50	−100	−100	350	0	0	0	None
c	Yes	−100	100	200	0	0	100	200	0	L_5, L_6
d	Yes	50	−50	100	50	50	0	100	50	L_4, L_6, L_7

6.12 $\delta1 = 120$ ps; $\delta2 = 0$; $\delta3 = 120$ ps.

6.14 $\delta4 = 110$ ps; $\delta5 = 70$ ps; $\delta6 = 110$ ps; $\delta7 = 70$ ps.

Bibliography

[1]   D. Bailey and B. Benschneider, "Clocking Design and Analysis for a 600-MHz Alpha Microprocessor," *IEEE J. Solid-State Circuits*, vol. 33, no. 11, pp. 1627–1633, Nov. 1998.

[2]   D. Bearden et al., "A 133 MHz 64b Four-Issue CMOS Microprocessor," *ISSCC Dig. Tech. Papers*, pp. 174–175, Feb. 1995.

[3]   B. Benschneider et al., "A 1GHz Alpha Microprocessor," *ISSCC Dig. Tech. Papers*, pp. 86–87, Feb. 2000.

[4]   L. Boonstra, C. Lambrechtse, and R. Salters, "A 4096-b One-Transistor per Bit Random-Access Memory with Internal Timing and Low Dissipation," *IEEE J. Solid-State Circuits*, vol. SC-8, no. 5, pp. 305–310, Oct. 1973.

[5]   W. Bowhill et al., "A 300 MHz 64 b Quad-Issue CMOS Microprocessor," *ISSCC Dig. Tech. Papers*, pp. 182–183, Feb. 1995.

[6]   W. Bowhill et al., "Circuit Implementation of a 300-MHz 64-Bit Second-Generation CMOS Alpha CPU," *Digital Technology Journal*, vol. 7, no. 1, pp. 100–119, 1995.

[7]   T. Burks, K. Sakallah, and T. Mudge, "Critical Paths in Circuits with Level-Sensitive Latches," *IEEE Trans. VLSI Sys.*, vol. 3, no. 2, pp. 273–291, June 1995.

[8]   A. Champernowne, L. Bushard, J. Rusterholtz, and J. Schomburg, "Latch-to-Latch Timing Rules," *IEEE Trans. Comput.*, vol. 39, no. 6, pp. 798–808, June 1990.

[9]   T. Chao, Y. Hsu, J. Ho, and A. Kahng, "Zero Skew Clock Routing with Minimum Wirelength," *IEEE Trans. Circuits Syst.-II*, vol. 39, no. 11, pp. 799–814, Nov. 1992.

[10]   D. Chapiro, *Globally-Asynchronous Locally Synchronous Systems*, Ph.D. dissertation, CS Department, Stanford University, Stanford, CA, May 1984.

[11] T. Chappell, B. Chappell et al., "A 2-ns Cycle, 3.8-ns Access 512-kb CMOS ECL SRAM with a Fully Pipelined Architecture," *IEEE J. Solid-State Circuits*, vol. 26, no. 11, pp.1577–1585, Nov. 1991.

[12] K. Chu and D. Pulfrey, "Design Procedures for Differential Cascode Voltage Switch Circuits," *IEEE J. Solid-State Circuits*, vol. SC-21, no. 6, pp. 1082–1087, Dec. 1986.

[13] R. Colwell and R. Steck, "A 0.6μm BiCMOS Processor with Dynamic Execution," *ISSCC Dig. Tech. Papers*, pp. 176–177, Feb. 1995.

[14] W. Dally and J. Poulton, *Digital Systems Engineering*, New York: Cambridge University Press, 1998.

[15] S. DasGupta, E. Eichelberger, and T. Williams, "LSI Chip Design for Testability," *ISSCC Dig. Tech. Papers*, pp. 216–217, Feb. 1978.

[16] D. Dobberpuhl et al., "A 200 MHz 64 b Dual-Issue CMOS Microprocessor," *IEEE J. Solid-State Circuits*, vol. 27, no. 11, pp. 1555–1567, Nov. 1992.

[17] D. Dobberpuhl et al., "A 200-MHz 64-Bit Dual-issue CMOS Microprocessor," *Digital Technology Journal*, vol. 4, no. 4, pp. 35–50, 1992.

[18] H. Fair and D. Bailey, "Clocking Design and Analysis for a 600 MHz Alpha Microprocessor," *ISSCC Dig. Tech. Papers*, pp. 398–399, Feb. 1998.

[19] E. Friedman, ed., *Clock Distribution Networks in VLSI Circuits and Systems*, New York: IEEE Press, 1995.

[20] N. Gaddis and J. Lotz, "A Quad-Issue Out-of-Order RISC CPU," *ISSCC Dig. Tech. Papers*, pp. 210–211, Feb. 1996.

[21] N. Gaddis and J. Lotz, "A 64-b Quad-Issue CMOS RISC Microprocessor," *IEEE J. Solid-State Circuits*, vol. 31, no. 11, pp. 1697–1702, Nov. 1996.

[22] B. Gieseke et al., "A 600 MHz Superscalar RISC Microprocessor with Out-of-Order Execution," *ISSCC Dig. Tech. Papers*, pp. 176–177, Feb. 1997.

[23] R. Ginosar and R. Kol, "Adaptive Synchronization," *Proc. Intl. Conf. Comp. Design*, pp. 188–189, Oct. 1998.

[24] N. Gonclaves and H. De Man, "NORA: A Racefree Dynamic CMOS Technique for Pipelined Logic Structures," *IEEE J. Solid-State Circuits*, vol. SC-18, no. 3, pp. 261–266, June 1983.

[25] R. Gonzalez and M. Horowitz, "Energy Dissipation in General Purpose Microprocessors," *IEEE J. Solid-State Circuits*, vol. 31, no. 9, pp. 1277–1284, Sept. 1996.

[26] P. Gronowski and B. Bowhill, "Dynamic Logic and Latches—Part II," *Proc. VLSI Circuits Workshop*, VLSI Circuits Symp., June 1996.

[27] P. Gronowski et al., "A 433-MHz 64-b Quad-Issue RISC Microprocessor," *IEEE J. Solid-State Circuits*, vol. 31, no. 11, pp. 1687–1696, Nov. 1996.

[28] P. Gronowski et al., "High-Performance Microprocessor Design," *IEEE J. Solid-State Circuits*, vol. 33, no. 5, pp. 676–686, May 1998.

[29] A. Hall, *Synthesis of Double Rank Sequential Circuits*, Tech. Report #53, EE Digital Systems Lab, Princeton University, Dec. 1966.

[30] D. Harris and M. Horowitz, "Skew-Tolerant Domino Circuits," *IEEE J. Solid-State Circuits*, vol. 32, no. 1, pp. 1702–1711, Nov. 1997.

[31] D. Harris, S. Oberman, and M. Horowitz, "SRT Division Architectures and Implementations," *Proc. 13th IEEE Symposium on Computer Arithmetic*, July 1997.

[32] R. Heald et al., "Implementation of a 3rd-Generation SPARC V9 64b Microprocessor," *ISSCC Dig. Tech. Papers,* pp. 412–413, Feb. 2000.

[33] C. Heikes, "A 4.5 mm^2 Multiplier Array for a 200MFLOP Pipelined Coprocessor," *ISSCC Dig. Tech. Papers*, pp. 290–291, Feb. 1994.

[34] C. Heikes and G. Colon-Bonet, "A Dual Floating Point Coprocessor with an FMAC Architecture," *ISSCC Dig. Tech. Papers*, pp. 354–355, Feb. 1996.

[35] L. Heller, W. Griffin, J. Davis, and N. Thoma, "Cascode Voltage Switch Logic: A Differential CMOS Logic Family," *ISSCC Dig. Tech. Papers*, pp. 16–17, Feb. 1984.

[36] J. Hennessy and D. Patterson, *Computer Architecture: A Quantitative Approach*, San Francisco: Morgan Kaufmann, Chapter 1, 1999.

[37] R. Hitchcock, "Timing Verification and Timing Analysis Program," *25 Years of Electronic Design Automation*, New York: IEEE/ACM, 1988.

[38] P. Hofstee et al., "A 1 GHz Single-Issue 64b PowerPC Processor," *ISSCC Dig. Tech. Papers,* pp. 92–93, Feb. 2000.

[39] M. Horowitz, "High Frequency Clock Distribution," *Proc. VLSI Circuits Workshop*, VLSI Circuits Symp., June 1996.

[40] *IBM J. Research & Dev.*, special issue on soft errors, vol. 40, no. 1, Jan. 1996.

[41] Intel Corporation, *Opportunistic Time-Borrowing Domino Logic*, U.S. Patent #5,517,136, May 14, 1996.

[42] Intel Corporation, "Intel Microprocessor Quick Reference Guide," courtesy of Intel Museum, Santa Clara, CA, 1997.

[43] Intel Corporation, *Pulsed Domino Latches*, U.S. Patent #5,880,608, March 9, 1999.

[44] A. Ishii, C. Leiserson, and M. Papaefthymiou, "Optimizing Two-Phase, Level-Clocked Circuitry," *J. ACM*, vol. 44, no. 1, pp. 148–199, Jan. 1997.

[45] N. Jouppi, *Timing Verification and Performance Improvement of MOS VLSI Designs*, Ph.D. thesis, Stanford University, 1984.

[46] V. von Kaenel et al., "A 600 MHz CMOS PLL Microprocessor Clock Generator with a 1.2 GHz VCO," *ISSCC Dig. Tech. Papers*, pp. 396–397, Feb. 1998.

[47] F. Klass, "Semi-Dynamic and Dynamic Flip-Flops with Embedded Logic," *Symposium on VLSI Circuits Dig. Tech. Papers*, pp. 108–109, June 1998.

[48] F. Klass et al., "A New Family of Semidynamic and Dynamic Flip-Flops with Embedded Logic for High-Performance Processors," *IEEE J. Solid-State Circuits*, vol. 34, no. 5, pp. 712–716, May 1999.

[49] R. Krambeck, C. Lee, and H. Law, "High-Speed Compact Circuits with CMOS," *IEEE J. Solid-State Circuits*, vol. SC-17, no. 3, pp. 614–619, 1982.

[50] J. Kuskin et al., "The Stanford FLASH Multiprocessor," *Proc. Intl. Symp. Comp. Arch.*, pp. 302–313, Apr. 1994.

[51] P. Larsson and C. Svensson, "Noise in Digital Dynamic CMOS Circuits," *IEEE J. Solid-State Circuits*, vol. 29, no. 6, June 1994.

[52] L. Lev, "Signal and Power Network Integrity," *Proc. VLSI Circuits Workshop*, VLSI Circuits Symp., June 1996.

[53] L. Lev et al., "A 64-b Microprocessor with Multimedia Support," *IEEE J. Solid-State Circuits*, vol. 30, no. 11, Nov. 1995.

[54] J. Lotz et al., "A Quad-Issue Out-of-Order RISC CPU," *ISSCC Dig. Tech. Papers*, pp. 210–211, Feb. 1996.

[55] M. Matsui et al., "A 200-MHz 13 mm^2 2-D DCT Macrocell Using Sense-Amplifier Pipeline Flip-Flop scheme," *IEEE J. Solid-State Circuits*, vol. 29, no. 12, pp. 1482–1491, Dec. 1994.

[56] C. Mead and L. Conway, *Introduction to VLSI Systems*, Reading, MA: Addison-Wesley, 1980.

[57] *Microprocessor Report*, Sebastopol, CA: MicroDesign Resources, 1995–1998.

[58] J. Montanaro et al., "A 160-MHz, 32-b, 0.5-W CMOS RISC Microprocessor," *IEEE J. Solid-State Circuits*, vol. 31, no. 11, pp. 1703–1714, Nov. 1996.

[59] G. Moore, "Cramming More Components onto Integrated Circuits," *Electronics*, pp. 114–117, Apr. 1965.

[60] A. Mukherjee, *Introduction to nMOS and CMOS VLSI Systems Design*, Englewood Cliffs, NJ: Prentice-Hall, 1986.

[61] D. Noice, *A Clocking Discipline for Two-Phase Digital Integrated Circuits*, Stanford University Technical Report, Jan. 1983.

[62] K. Nowka and T. Galambos, "Circuit Design Techniques for a Gigahertz Integer Microprocessor," *Proc. Intl. Conf. Comp. Design*, pp. 11–16, Oct. 1998.

[63] J. Ousterhout, "A Switch-Level Timing Verifier for Digital MOS VLSI," *IEEE Trans. Computer-Aided Design*, vol. CAD-4, no. 3, pp. 336–349, July 1985.

[64] H. Partovi et al., "Flow-Through Latch and Edge-Triggered Flip-Flop Hybrid Elements," *ISSCC Dig. Tech. Papers*, pp. 138–139, Feb. 1996.

[65] W. Penny and L. Lau, *MOS Integrated Circuits: Theory, Fabrication, Design and Systems Applications of MOS LSI*, New York: Van Nostrand, Reinhold, Chapter 5, 1973.

[66] J. Rabaey, *Digital Integrated Circuits*, Upper Saddle River, NJ: Prentice-Hall, 1996.

[67] B. Razavi, ed., *Monolithic Phase-Locked Loops and Clock Recovery Circuits*, New York: IEEE Press, 1996.

[68] P. Restle and A. Deutsch, "Designing the Best Clock Distribution Network," *Proc. VLSI Symp.*, pp. 2–5, June 1998.

[69] J. Rubinstein, P. Penfield, and M. Horowitz, "Signal Delay in RC Tree Networks," *IEEE Trans. Computer-Aided Design*, vol. CAD-2, no. 3, pp. 202–211, July 1983.

[70] K. Sakallah, T. Mudge, and O. Olukotun, "Analysis and Design of Latch-Controlled Synchronous Digital Circuits," *IEEE Trans. Computer-Aided Design*, vol. 11, no. 3, pp. 322–333, Mar. 1992.

[71] Semiconductor Industry Association, *The National Technology Roadmap for Semiconductors*, Austin, TX: SEMATECH, 1997. (*notes.sematech.org/97melec.htm*)

[72] Semiconductor Industry Association, *International Technology Roadmap for Semiconductors*, 1999. (*www.itrs.net/199 SIA Roadmap/Home.htm*)

[73] N. Shenoy, *Timing Issues in Sequential Circuits*, Ph.D. dissertation, University of California, Berkeley, 1993.

[74] N. Shenoy, R. Brayton, and A. Sangiovanni-Vincentelli, "A Pseudo-Polynomial Algorithm for Verification of Clocking Schemes," *Tau*, 92, 1992.

[75] K. Shepard et al., "Design Methodology for the S/390 Parallel Enterprise Server G4 Microprocessors," *IBM J. Research and Development*, vol. 41, no. 4–5, pp. 515–547, July–Sept. 1997.

[76] K. Shepard, V. Narayanan, and R. Rose, "Harmony: Static Noise Analysis of Deep Submicron Digital Integrated Circuits," *IEEE Trans. Computer-Aided Design*, vol. 18, no. 8, pp. 1136–1150, Aug. 1999.

[77] M. Shoji, "Electrical Design of BELLMAC-32A Microprocessor," *Proc. IEEE Int'l Conf. Circuits and Computers*, pp. 112–115, Sept. 1982.

[78] M. Shoji, "Elimination of Process-Dependent Clock Skew in CMOS VLSI," *IEEE J. Solid-State Circuits*, vol. SC-21, no. 5, pp. 875–880, Oct. 1986.

[79] M. Shoji, *High-Performance CMOS Circuits*, Englewood Cliffs, NJ: Prentice-Hall, 1988.

[80] G. Singer and S. Rusu, "The First 1A-64 Microprocessor: A Design for Highly Parallel Execution," *ISSCC Dig. Tech. Papers*, pp. 422–423, Feb. 2000.

[81] V. Stojanovic and V. Oklobdzija, "Comparative Analysis of Master-Slave Latches and Flip-Flops for High-Performance and Low-Power Systems," *IEEE J. Solid-State Circuits*, vol. 34, no. 4, pp. 536–548, Apr. 1999.

[82] I. Sutherland, R. Sproull, and D. Harris, *Logical Effort*, San Francisco, CA: Morgan Kaufmann, 1999.

[83] T. Szymanski, "LEADOUT: A Static Timing Analyzer for MOS Circuits," in *ICCAD-86 Dig. Tech. Papers*, pp. 130–133, 1986.

[84] T. Szymanski, "Computing Optimal Clock Schedules," *Proc. 29th Design Automation Conf.*, pp. 399–404, 1992.

[85] T. Szymanski and N. Shenoy, "Verifying clock schedules," *ICCAD Dig. Tech. Papers*, pp. 124–131, Nov. 1992.

[86] O. Takahashi, N. Aoki, J. Silberman, and S. Dhong, "A 1-GHz Logic Circuit Family with Sense Amplifiers," *IEEE J. Solid-State Circuits*, vol. 34, no. 5, pp. 616–622, May 1999.

[87] T. Thorp, G. Yee, and C. Sechen, "Domino Logic Synthesis Using Complex Gates," *Proc. Intl. Conf. Computer-Aided Design*, Nov. 1998.

[88] R. Tsay, "An Exact Zero-Skew Clock Routing Algorithm," *IEEE Trans. Computer-Aided Design*, vol. 12, no. 2, pp. 242–249, Feb. 1993.

[89] S. Unger and C. Tan, "Clocking Schemes for High-Speed Digital Systems," *IEEE Trans. Comput.*, vol. C-35, no. 10, pp. 880–895, Oct. 1986.

[90] N. Vasseghi et al., "200 MHz Superscalar RISC Processor," *IEEE J. Solid-State Circuits*, vol. 31, no. 11, pp. 1675–1686, Nov. 1996.

[91] A. Vittal and M. Marek-Sadowska, "Crosstalk Reduction for VLSI," *IEEE Transactions on CAD*, vol. 16, no. 3, pp. 290–298, March 1997.

[92] N. Weste and K. Eshraghian, *Principles of CMOS VLSI Design*, Reading, MA: Addison-Wesley, p. 351, 1993.

[93] T. Williams, *Self-Timed Rings and Their Application to Division*, Ph.D. dissertation, EE Department, Stanford University, Stanford, CA, May 1991.

[94] T. Williams and M. Horowitz, "A Zero-Overhead Self-Timed 160-ns 54-b CMOS Divider," *IEEE J. Solid-State Circuits*, vol. 26, no. 11, pp. 1651–1661, Nov. 1991.

[95] Y. Ye, *Interior Point Algorithms: Theory and Analysis,* New York: Wiley, 1997.

[96] J. Yuan and C. Svensson, "High Speed CMOS Circuit Technique," *IEEE J. Solid-State Circuits*, vol. 24, no. 1, pp. 62–69, Feb. 1989.

[97] J. Yuan and C. Svensson, "New Single-Clock SMOS Latches and Flipflops with Improved Speed and Power Savings," *IEEE J. Solid-State Circuits*, vol. 32, no. 1, pp. 62–69, Jan. 1997.

[98] J. Yuan, C. Svensson, and P. Larsson, "New Domino Logic Precharged by Clock and Data," *Electronics Letters*, vol. 29, no. 25, pp. 2188–2189, Dec. 1993.

Index

About the Author

David Harris is currently an Assistant Professor of Engineering at Harvey Mudd College. He received his Ph.D. in 1999 from Stanford University on skew-tolerant circuit design. Since receiving his M. Eng. from MIT in 1994, he has consulted and taught in the field of high-speed CMOS circuit design at Sun Microsystems, Intel Corporation, HAL Computer, and Evans & Sutherland. In addition, he has taught circuit design at the U.C. Berkeley Extension and Stanford University. When David is not building chips or teaching VLSI to freshmen, he can often be found mountaineering or flying a Cessna.